Second Edition

Statistics for the Biological Sciences

WILLIAM C. SCHEFLER

State University College at Buffalo

Second Edition

Statistics for the Biological Sciences

**ADDISON-WESLEY
PUBLISHING COMPANY**

Reading, Massachusetts
Menlo Park, California · London · Amsterdam
Don Mills, Ontario · Sydney

ISBN 0-201-07500-8
ABCDEFGHIJK-MA-798

To Wilma

For over three decades of support and understanding

Preface
to the Second Edition

In a very real sense, the first edition of a textbook is only a first draft. The realities of the classroom then form the crucible in which the second edition is forged. Thus, the content of the second edition of *Statistics for the Biological Sciences* in part reflects the author's experiences with the book in his own course and in part reflects the criticisms and suggestions contributed by others who use it in their classes.

The preparation of the second edition has involved extensive rewriting in certain areas. For example, the chapter on correlation and regression now makes a careful distinction between these two procedures and between the kinds of experimental situations to which each can legitimately be applied. Also, a reorganization and rewriting of the material on analysis of variance, with the addition of a multiple-range test, have improved the presentation of this very important procedure. The chapter on non-parametric statistics has been shifted to the end of the book and has been rounded out with the addition of the Kruskal–Wallis and Friedman χ_r^2 tests. The chapter on probability has been revised to include fewer applications to coins and dice and more to biology.

Chapter 1 is an entirely new chapter which includes an introductory discussion of basic principles of experimental design and the analysis of data. These same principles, however, continue to be stressed throughout the book as they arise.

Finally, the text has been carefully checked for errors, and statistical symbols have been changed where necessary to conform to generally accepted practice.

All revisions and additions have been made without losing sight of the fact that whatever success *Statistics for the Biological Sciences* has enjoyed has been largely due to its simplicity of style and content. Thus, the original style has been maintained and continues to reflect the conviction that it is both desirable and possible to introduce useful principles of statistics without the rigorous mathematical treatment and summation notation that tend to reduce many biology students to a state of helpless confusion.

Two things related to the reader's background are assumed. First, he or she should be able to remember a fair amount of beginning high-school algebra. Second, a reasonably good background in biology is helpful in following the kind of examples that are used.

The preparation of this book was guided by the opinion that applied statistics should not be just another mathematics course. It involves a way of thinking, and it should be part of the background of every modern biologist. It has its own special body of principles which can be approached intuitively; and an understanding of these principles need not depend on a special aptitude for mathematics.

Biometrics should not be regarded simply as an "extra" to which the student may be exposed if it happens to fit into his or her program. We go to great lengths to ensure that the biology major becomes familiar with microscopes, microtomes, centrifuges, electrophoresis apparatus, and other standard tools of the trade. Concepts that deal with experimental design and analysis of data constitute another indispensable tool of the biologist, and it is every bit as important that he or she be as familiar with these concepts as with the proper operation of a microscope.

Like the first edition, the second edition of *Statistics for the Biological Sciences* does not pretend to be a scholarly work in biostatistics, nor does it try to make statisticians out of biologists. It does attempt to provide biology students, teachers, research biologists, and others with a substantial amount of basic statistical theory that is logically rather than mathematically derived. Most of all, its purpose is to prepare them for their most critical task—to talk intelligently with statisticians.

I am grateful to the various authors and publishers who gave me permission to reprint the tables that appear in the Appendix of this book. The source for each table is given in the footnote below the table.

I am indebted to the Literary Executor of the late Sir Ronald A. Fisher, F.R.S., to Dr. Frank Yates, F.R.S., and to Oliver and Boyd, Ltd., Edinburgh, for permission to reprint the tables of the *t*-distribution and the *r*-to-*Z* transformation from their book, *Statistical Tables for Biological, Agricultural and Medical Research*.

Buffalo, N.Y.   W.C.S.
December 1978

Contents

CHAPTER EIGHT

ANALYSIS OF VARIANCE—PART I

CHAPTER NINE

ANALYSIS OF VARIANCE—PART II

CHAPTER TEN

CORRELATION AND REGRESSION

CHAPTER ONE

Statistics—A Tool of Research

1.1 INTRODUCTION

A working knowledge of statistics and experimental design is an indispensable part of the background of the modern biologist who wishes to carry on serious research. Furthermore, even those who simply want to intelligently read articles in biological journals cannot do so without more than just a passing acquaintance with statistical procedures. For example, what does it mean when a scientist reports that he or she used a "randomized block design" and found the results "significant beyond the 0.01 level?" Or what is implied by the phrase, "the mean plus-or-minus two standard errors?"

It can easily be argued that a knowledge of biostatistics, or *biometrics*, is as important to the modern biologist as an ability to use a pH meter or to understand a complex chemical formula. On the other hand, it would be unreasonable to expect the biologist to design and build pH meters, amino-acid analyzers, and other complicated tools of the trade in order to *use* such things intelligently. Fortunately, it is equally unnecessary to be a mathematician in order to acquire a practical knowledge of statistics and its applications to experimental work. The mathematicians have obligingly "designed and built" the tools of statistics, and the often less mathematically inclined biologist can easily learn to apply these tools to his or her own research. In fact, it should be stressed that the concepts of applied statistics constitute a special body of knowledge, and a course in biostatistics should by no means be regarded as just another mathematics course. The study of applied statistics and experimental design emphasizes methods of reasoning, and more so-called "critical thinking" and "scientific method" are found in a good applied statistics course than are found in a score of traditional biology courses.

This book assumes no mathematics background beyond an ability to perform simple algebraic manipulations of the kind usually learned in early secondary school. On the other hand, because applied statistics and principles of experimental design are indispensable tools of the scientist, they may be studied with profit by the more mathematically inclined biologist as well as by those who flee in terror at the mention of the word *calculus*.

1.2 ORIGIN OF STATISTICS

Statistics originally involved facts and figures gathered for various purposes by governments. The periodic census is a modern example of this activity. It thus appears that the word *statistics* is derived from the *state* or government.

This type of statistics is often referred to as *descriptive*. Items such as the average annual number of deaths on the highways over the past ten years, or a count of shore birds taken during the month of April, are examples of descriptive statistics. In such cases we are simply describing a situation in terms of facts and figures, or *data*. As biologists, we are interested in descriptive statistics (data), but in most research situations we are concerned about the procedures of *inferential statistics*. Inferential statistics is a major tool of the research biologist and we will look at it in considerable detail as we go through this book.

The development of biostatistics, or *biometrics*, was considerably influenced by the work of researchers in the field of agriculture. In fact, some biometrics books still refer to experimental subjects as "plots" even when they are rats, mice, or human subjects. Also, the practice of dividing treatments into "levels" comes from experiments in which different levels of fertilizer, etc., were used.

1.3 THE MISUSE OF STATISTICS

There is an old saying that "figures don't lie but liars do figure." In fact, the British Prime Minister Disraeli was so disenchanted with our favorite subject that he classified falsehoods in increasing order of shamefulness as "lies, damned lies, and statistics!"

Like many other useful and effective tools, statistics certainly can be misused. For instance, there are numerous examples of deliberate misuse of statistics by those who use their statistical skills to serve special interests by cleverly distorting and misrepresenting data. Exaggerated or downright fraudulent advertising claims are all too familiar examples of such statistical quackery. The politician's campaign oratory is often laced with deliberately one-sided and distorted "statistical evidence" that is designed to further his or her own political fortunes. The incumbent may boast that more people are working today than at any other time in history. The challenger replies that more people are unemployed today than at any other time in history. Both may be right; there are *more people* today than at any other time in history. One simply chooses to tell us that the glass is half full while the other tells us that it is half empty. For an entertaining exposé of this seamier side of our subject, the reader is referred to Huff, *How to Lie with Statistics.*

Of course, the misuse of statistics is not always the work of a statistical crook. Sometimes, well-meaning and otherwise competent researchers get into trouble because they either know too little about statistical concepts and procedures or know too much that is not so. Sometimes, too, the "big name syndrome" helps to perpetuate poor statistical techniques. Young scientists, especially, tend to be overawed by people with established reputations; the statistical sins found in Dr. Blank's publications are therefore often repeated because Dr. Blank cannot be wrong.

Most of us grew up with the traditional belief that scientists are open-minded, objective, selfless, honest, self-critical, and, above all, filled with a religious zeal to pursue the "truth" no matter where the trail shall lead. Fortunately, we probably can be confident that most of this is true most of the time. On the other hand, to believe that all of it is true all of the time is to assume that scientists are either less than human or more than human. Only the naive could really mistake the scientist's traditional white coat for a cloak of immunity to the human weaknesses common to us all. It is doubtful that every researcher is so immersed in a "search for the truth" as to be deaf to the call of fame and fortune. And what biologist would not dare to dream of a place alongside the immortal Darwin? In many situations there are pressures to produce, to publish, to win larger and larger grants, because these are the foundations upon which a reputation is built. Considering all this, it is not surprising that researchers have been known to "cut corners," to publish prematurely, and to make claims not fully justified by their data.

The importance of statistics is not diminished because it is sometimes misused, either by sharp operators or by bumbling if well-meaning researchers. One may as well condemn such a useful and life-saving tool as the scalpel because it is occasionally wielded by an incompetent or unscrupulous surgeon. Like the scalpel, the statistical test has no mind of its own, and its usefulness depends on its user.

1.4 THE NATURE OF BIOLOGICAL RESEARCH

It would not be appropriate to define biological research solely in terms of certain specific and restricted activities. The legitimate research activities of biologists cover an extremely broad spectrum that blends almost imperceptibly with other sciences such as chemistry, geology, physics, and psychology. These activities range from the carefully controlled laboratory experiments of the physiologist to the floristic studies of the plant taxonomist.

In a general way, biological research involves carefully planned and executed activities that are performed with the hope of adding a (usually) limited amount of knowledge in a (usually) limited area. While some still like to think that the researcher formulates a "hypothesis" and then designs an experiment to "prove" it, modern researchers tend to think in terms of "asking questions" relative to a specific problem. They will then seek information concerning those questions through a specific research activity.

For example, a pharmacologist may question whether or not a recently developed vaccine offers effective protection against a certain infectious disease. On the other hand, a plant taxonomist may be concerned with conducting a survey of the plant species found in a certain geographic area. Meanwhile, a geneticist may try to determine the genetic mechanism that controls the inheritance pattern of a certain trait. Finally, an ecologist wishes to know how a thermal effluent from a nuclear plant is changing the ecosystem of a lake.

The foregoing represent only a few of the many diverse questions that biologists in different fields may "ask of Nature." It follows that legitimate and important biological research can take many forms; each represents a special problem that requires special expertise and special research methods.

1.5 THE IMPORTANCE OF EXPERIMENTAL DESIGN

The use of statistics as a tool of research cannot be separated from the general planning of the research project. If a research project is to yield data that are to be statistically treated, then an appropriate statistical test *must* be an integral part of the total design. Nothing adds more to the woes of the statistician than the naive researcher who gathers data with the blithe assumption that a workable statistical test will automatically be available.

The experimental approach to seeking information concerning a biological problem must involve careful thought and advance planning. In other words, although it may seem too fundamental to even mention, *a research project must be designed and planned before it is carried out.* Yet, obvious as this may seem, statisticians are only too familiar with the researcher who brings in reams of data that were gathered in a haphazard fashion, often without any clear notion of why the data had been obtained in the first place! In such cases it is often the statistician's sad duty to tell the researcher that his or her efforts were largely wasted because there is no legitimate way to handle the data.

When we carefully plan a research project we can do some very important things such as (1) balance the desired sensitivity of a test against cost (a very important consideration!), (2) determine what variables are present and how they can be controlled, (3) determine how precise our measurements should be, (4) decide what sampling methods shall be used, and (5) deal with the problem of sample size.

Also, we can determine beforehand what statistical tests are available to analyze our data while taking into consideration the measurement scale used, the experimental design, normality of the data, and a variety of other factors. Therefore, we will say it again: *The research must be planned before it is begun, and the plan must include an appropriate statistical treatment if statistical analysis is indicated.*

Ideally, the research project is the dog and the statistical test is the tail. The tail, however, is an integral part of the dog and was not added on as an afterthought. Still, the tail should not be allowed to wag the dog. In other words, statistical tests are an aid to thinking and should not be allowed to dictate how the researcher should think. Statistics is a servant, not a master.

Because of their extreme importance, fundamentals of experimental design will be stressed at appropriate places throughout this book. Also, in a later chapter, we will consider some of the more complex designs that are highly useful to the biologist. We will begin in the next section by exploring the concept of the so-called "true experiment."

1.6 THE EXPERIMENT

Suppose that you wish to find out what relationship, if any, exists between the volume of a gas and its temperature. As an approach to this problem, you decide to subject the gas to a variety of temperatures and to record the gas volume observed at each of the temperatures chosen. In this situation, *you* are deciding which temperatures to use. The variation of temperature is under the control of the experimenter; temperature is therefore called the *independent variable*. On the other hand, you have no control over the gas volume that may result from a given temperature value. This will be "decided by nature" in accordance with whatever relationship exists between temperature and volume. Volume is therefore called the *dependent variable* because the variation in volume *depends* upon the variation in temperature. In other words, the independent variable is changed at the will of the experimenter, while the dependent variable changes according to the values of the independent variable.

This of course assumes that some kind of "natural" relationship between the dependent and independent variables actually exists. In our

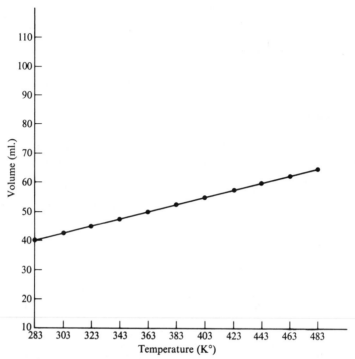

Fig. 1.1 Relationship between temperature (K°) and volume of an ideal gas. Temperature is the independent variable and volume is the dependent variable.

case in point, Fig. 1.1 shows that our experiment demonstrates a direct proportional relationship between volume and temperature. You have most likely already recognized this as the classic laboratory exercise that illustrates Charles' Law.

At this point we must be careful not to draw hasty conclusions unless we are reasonably certain that temperature was the only independent variable affecting the dependent variable (volume). It would have been a reasonable assumption, for example, that changes in pressure would have also brought about changes in gas volume. In performing the experiment, therefore, pressure would have to be kept constant while temperature was varied. Otherwise, we could never be sure whether the independent variable *temperature*, the independent variable *pressure*, or a combination of both was causing the volume to vary in a particular way.

To further illustrate this with a silly example, suppose we were to compare the effects of two dog foods on growth. In this situation we are manipulating an independent variable (nutrition) and observing a dependent variable (growth). We now carry on an experiment in which we feed

"Barks and Bites" dog food to a group of German shepherd puppies and "Brand X" to a group of Pekinese puppies. At the end of one year we evaluate the results and "Barks and Bites" is an easy winner. We now use our results to make a commercial for "Barks and Bites" and wonder why it seems to convince very few people that we have "scientific" evidence that "Barks and Bites" grows bigger dogs.

It is of course quite obvious that our very poorly planned experiment did nothing to separate nutritional effects from genetic effects. The nutritional and genetic factors are both independent variables and both have an obvious role in determining how large a dog will be after one year's growth. We can therefore draw no conclusion concerning the nutritional value of "Barks and Bites." When two or more independent variables act upon a dependent variable in this mixed-up fashion, we say that they are *confounded*. This confounding of independent variables is one of the bugaboos of the researcher; the identification and control of extraneous independent variables is therefore an extremely important aspect of experimental design.

At this point we are ready to attempt a definition of the term *experiment*. In formal terms, the "true experiment" is a situation in which the experimenter manipulates a specific independent variable and observes the effects of this manipulation on a specific dependent variable while keeping other independent variables constant. This is a formal and oversimplified description of an experiment, and we shall see numerous variations on this theme as we go through later chapters.

It would be wrong to assume that the "true experiment" that we have described so formally above represents the only valid kind of research in biology. As we pointed out earlier, the biological sciences and their problems are far too broad in scope to permit the restriction of all biological research projects to rigidly controlled laboratory experiments. The plant taxonomist is not performing a "true experiment" by walking over an area to survey the plant species. Neither, for that matter, are the researchers who take samples from a drifting boat and contend with a host of variables as they look at the effects of pollution on a large lake ecosystem. On the other hand, the plant taxonomist confined to a laboratory will never know what species exist in a designated area, and certainly no one has ever obtained even a reasonably good picture of a lake ecosystem by staying on shore.

1.7 TWO-GROUP DESIGNS

At this point we are ready to follow the basic reasoning that underlies simple two-group experimental designs. In later portions of this book we will learn how to *statistically analyze* the data obtained from these and other designs by a procedure called *hypothesis testing*.

Suppose that we wish to determine whether a newly developed drug significantly elevates pulse rate in human subjects. Suppose, further, that we begin by selecting an adult female and measuring her pulse rate prior to the injection of the drug and again fifteen minutes after the drug is injected. This yields the following data:

Before	After
75	85

Now, just as we are about to set up a cheer for our drug, someone points out that almost anyone's pulse rate will increase when someone approaches with a needle! At that point we have to admit that the injection itself is an independent variable which is confounded with the effects of the drug. This brings up the need for a *control*, so we change our plans a bit and search around for another subject to act as a control. It so happens that for this purpose we select a male. This time we inject the female with the drug and the male receives a *placebo*, which is an inert substance having no effect on pulse rate (we hope!), and after a fifteen-minute interval we obtain the following data:

Female (drug)	Male (placebo)
85	78

As we rush to publish our results, someone reminds us that females are likely to have higher pulse rates than males. We consult the literature and find that females, as a group, do indeed tend to have higher pulse rates than males. Thus we have once again confounded two independent variables, both of which could affect the dependent variable (pulse rate). In other words, the effects of the drug and sex factors are so thoroughly confused that we cannot separate the effects of one from the other. Obviously, no conclusions can be drawn concerning the effect of the drug alone.

We now improve things by substituting an adult male for the female. Having learned of the danger of confounding, we are more cautious this time and are careful to choose a control subject who is as alike in age and size to the experimental subject as possible. This time we obtain

Drug (male)	Placebo (male)
86	78

Once again, the troublemaker spoils the celebration by asking another irritating question: "Is it not true that a considerable variation in pulse rate exists among human subjects, even when they are of the same sex, size, and age? Therefore, if the procedure were repeated, using two different subjects, is it not possible that the results could even be reversed?"

Having been forced to think once again, we now have to admit that there is considerable variability among organisms of the same species. For example, given a group of adult human males and given measuring devices of sufficient sensitivity, we would rarely find any two individuals with *exactly* the same height, weight, blood pressure, etc. This natural variation is called *error*, which in this case does not mean "mistake" but refers to the variability among subjects, which the experimenter cannot eliminate—and with which he or she must always contend.

We therefore decide to handle the problem of variability by repeating the experiment a number of times, using different subjects for each trial. Furthermore, the experiment will be repeated under the same conditions, or *replicated*. We will therefore need to have a certain number of *replicates* per treatment (drug and control).

Now, suppose that we have available 80 adult males of approximately the same size and same age. Our next step is to assign 40 of the subjects to the drug treatment and 40 to the placebo treatment. Remember that the drug and placebo treatments represent two "levels" of the independent variable (drug) and that we are trying to determine the relative effects of these two levels on the dependent variable (pulse rate). At this point we might be tempted to arbitrarily select 40 subjects for the drug group and let the remaining 40 subjects comprise the placebo group. Again, there is a catch. Is it not possible that we might consciously or subconsciously assign certain types to one group or the other? Is our "scientific objectivity" strong enough to overcome the temptation to assign a greater proportion of the nervous types to the drug group, leaving the calmer types for the placebo group? Considering that our dependent variable is pulse rate, the assignment of subjects in this way could definitely bias our results, helping to "prove" our contention that our drug elevates pulse rate!

Obviously, the choice of who goes to what group must somehow be taken out of our hands. Furthermore, the assignment of subjects must be taken out of anyone's hands, because one of the characteristics of the human mind seems to be a compulsive need to *make* experiments "turn out right." We will therefore turn to the table of random numbers (Table II in the appendix). This table may be used as follows:

1. Assign a number ranging from 1 to 80 to each of the 80 subjects in the group.

2. Drop a pencil point somewhere on the random numbers table. From the point of pencil contact, we proceed to underline succeeding two-digit groups which include the numbers 1–80. Thus we might have:

26 94 <u>03</u> <u>68</u> <u>58</u> <u>70</u> <u>29</u> <u>73</u> <u>41</u> <u>35</u> <u>53</u> <u>14</u> 03 <u>33</u> <u>40</u>

and so on. Note that once a number is underlined it is subsequently ignored, as in the case of the number 03.

3. After 40 subjects have been selected in this way, these subjects now constitute one group, and the remaining 40 comprise the second group. We now toss a coin to see which group receives the drug and which group is given the placebo.

Now we have established two groups, and although they are composed of different subjects, we are hopeful that our "complete randomization" procedure produced two groups that are not significantly different in terms of variation in response before treatment. In other words, any conclusions based on experimental results must include the assumption that both groups were drawn from the same statistical population. To use an example stated earlier, males and females are in different statistical populations—at least insofar as pulse rate is concerned.

We have just described a method that is sometimes called an "uncorrelated samples" comparison. There is another approach which will somewhat reduce error due to variability, provided we can assume that neither the drug nor the placebo will have a permanent effect on the subject's physiological state. If we are reasonably certain this assumption is valid, the same subjects can be used for both drug and placebo injections— at different times, of course. The placebo could be given first, the data collected, and then the drug could be given and its associated data collected. Or, with each subject, it might be randomly decided as to which shall be given first, the drug or the placebo. Yet another possibility would involve two separate experiments, using the same subjects, in which we reverse the order in which we administer the drug and the placebo. A comparison of results might then shed light on whether the order of administration of drug and placebo had any effect.

This procedure, known as a "matched-pair" design, has the obvious advantages inherent in using the same subject for both drug and placebo. Since the same subject is involved with both treatments, there is a high probability that extraneous factors affecting the variable are kept relatively constant. Naturally, it can be used only in cases where the subject is not permanently altered by either treatment.

Matched-pair designs sometimes involve a matching, or attempted matching, of two *different* subjects, which then makes up a replication. This is useful in diet experiments where pairs may be matched in terms of weight, sex, genetic background, etc. This procedure should be approached with caution, since it is not always easy to account for all the possible variables that could affect the validity of experimental results.

As we mentioned at the beginning of this section, our final step is to analyze the data collected from our experiment. In this case, for example, we want to know whether or not our drug appears to be effective. Or, to put it another way, is there a significant difference between the responses of the drug group and those of the placebo group? To get at this aspect of our research project requires a knowledge of hypothesis-testing procedures; we will attempt to build this knowledge in a step-by-step fashion throughout the remainder of this book.

1.8 THE USE OF COMPUTERS AND CALCULATORS

No one would deny that the computer is one of the most wonderful products of the twentieth-century. In only minutes, it can perform complex operations that a team of mathematicians would take months to complete. Computers have made it possible to send spaceships to the moon and mechanical explorers to the surface of Mars. To the researcher, the computer is a willing servant that saves untold hours of drudgery when significant amounts of data must be analyzed.

The automobile is also a wonderful machine that enables people to travel from place to place as they were never able to do with the horse and buggy. And yet we know that an automobile cannot think, it does not know whether to go right or left, it will just as willingly go into a tree as around it, and, unlike the horse, it can't even find its way home. In other words, it is only a machine and it does only what it is told to do by the human brain that is guiding it.

The computer, like the automobile, does only what it is told to do by the human brain that programs it. Contrary to tales of science fiction, the computer does not think. In fact, the computer is a very dumb machine that may send John C. Smith the bank statement that should go to John B. Smith, and it has been known to give a furious and frustrated John Jones a poor credit rating when the real culprit is Robert Jones. Regardless of make, cost, or type of program, the computer does only what it is told to do and does not make true value judgments concerning whether what it is told to do is right or wrong.

For example, the computer does not know (or care) whether the data that are fed into its programs have been produced from an experiment that was properly designed and carefully performed. Neither does it know (or care) whether the statistical test it is asked to perform is appropriate to the data or to the design. Consequently, the computer willingly produces an "answer" which the researcher all too often may accept as gospel—especially since it was provided by an expensive and sophisticated computer. To put it in terms often used by statisticians, if "garbage" is put into the computer it will produce "garbage." This "garbage in—garbage out" syndrome is a disease that affects the results obtained by biologists who are naive about the fundamentals of experimental design and the methods of inferential statistics.

It is certainly possible that, more than anything else, the *misuse* of computers is contributing to the statistical illiteracy found among too many researchers who are otherwise excellent biologists. It is perhaps too easy for the experimenter to gather data, take it to a computer center, and have an obliging computer subject the data to a statistical procedure with which the researcher has only a limited acquaintance and which may be entirely inappropriate to the project at hand. Indeed, one of the primary objectives of this book is to provide the researcher or future researcher with an understanding of the fundamentals of the more common designs and methods of statistical analysis. With such tools in hand, the computer becomes an unbelievably helpful instrument.

The desk calculator has undergone considerable changes in recent years. The old mechanical monster that clattered its way loudly toward a predictably early breakdown has been replaced by electronic models that are fast, silent, dependable, and increasingly sophisticated to the point where they are actually minicomputers. Desk calculators definitely have a place. Not all biological research projects produce data so extensive as to require the assistance of a computer. The busy researcher should stop and think twice before going off to a computer center with data that could easily be analyzed on a modern desk calculator in the time it takes to reach the computer center!

Whether data are analyzed by computer or desk calculator, the important thing to remember is that no machine, no matter how sophisticated, can substitute for the most powerful computer yet devised—the human brain. If you remember nothing else from this book, remember this—that the subjection of experimentally derived data to complicated statistics performed by expensive machines will *not* magically create valid results from a poorly designed and carelessly conducted experiment.

And now we have put off our look at statistics about as long as we decently can. In the next chapter we will begin to face the challenge of numbers. You will find that it is really an easy challenge to meet, and if you do so you will be rewarded with a new viewpoint of biology as well as with some very important fundamentals of research.

CHAPTER TWO

Data and Distributions

2.1 BIOLOGICAL MEASUREMENTS

In this and succeeding chapters we will spend much of our time discussing ways to handle data derived from biological research. First, however, we should understand that data fall into several categories; this is important because the choice of statistical treatment depends in part on the kind of data involved.

First, we have *measurement* data, both *discontinuous* and *continuous*. Discontinuous data, also called "discrete" or "meristic" data, are obtained by counting items such as the number of heads in 100 tosses of a coin, the number of bristles on a fruit fly, the number of eggs in a nest, and so on. When dealing with *discontinuous* data, we "jump" from one point to the next, with no measurements in between. In other words, you are not going to count 4.89 eggs in a nest or come up with 48.32 heads out of 100 tosses of a coin. On the other hand, *continuous* data involves measurements in centimeters, millimeters, microns, grams, and other standard units with which you are familiar. When we measure the length of a leaf, for example, we can come up with 4.28 centimeters because there is an infinite number of possible measurements from one end of the leaf to the other. Obviously, leaves do not come in lengths that jump from 2 to 3 to 4 to 5 centimeters with nothing in between.

A second category of data involves those that are measured on the *ordinal* scale. For example, suppose a researcher wishes to test the effect of a certain brain lesion on maternal behavior in mice. One of the criteria of maternal behavior is the quality of the nest that the female builds. How do we measure the "quality" of a nest? The experimenter might set up certain criteria that would make it possible to evaluate a nest as excellent, good, fair, or poor. These terms are now converted to the numbers 1, 2, 3, and 4, thus ranking the nests in decreasing order of quality. The experimenter now has numbers that can be subjected to a statistical test. This "ordinal" scale, however, does not have the strength of the measurement scale; as we shall see later on, it requires special statistical methods.

Finally, we often encounter biology data that fall in the category of the *nominal* scale. For example, we may look at a certain group of fruit flies and simply count (enumerate) those with vestigial wings and count those with normal wings. Or, given a sample of mice that have been given a vaccine and then challenged with the corresponding pathogen, we could count those that develop signs of the disease and count those who apparently remain healthy. These data are called *enumeration* data, and they require still another kind of statistical procedure.

A question that often arises has to do with the degree of precision with which measurement data are presented. To put it in another way, how close should we be to the "truth" when we say that an item weighs a certain number of grams? First of all, we must remember that continuous measurement data obtained in the laboratory or field are rarely, if ever, exact. How close a measurement comes to the actual value depends upon the sensitivity of available instruments and the skill of the experimenter. Furthermore, it is likely that, in the vast majority of cases, the *true* measurement will never be known and cannot be known with existing measuring instruments. Measurements are, therefore, to some degree only an approximation of the truth.

Suppose, for example, that we measure the length of a leaf, using a metric rule divided into centimeters and millimeters. Suppose further that the length of the leaf turns out to be between 6.6 and 6.7 centimeters. At this point we can *estimate* between 6.6 and 6.7 centimeters. If our estimate is, say, 0.4 millimeter, we can report our measurement as 6.64 centimeters. In this case, the first two figures are definite and the third figure is *estimated* from our measuring instrument. Our measurement therefore has three *significant figures*. Obviously, we could not obtain a measurement with four significant figures because we could not estimate anything beyond the third figure. Again, we cannot claim that the *true* measurement is 6.64 centimeters, because we are not certain of the third figure. All we can say is that the true length of the leaf falls between 6.635 and 6.645 centimeters. In a similar fashion, if we were to weigh an object on a balance that weighs to 0.001 gram, we could present a weight of 6.647 grams, but we could not go beyond four significant figures because the precision of our balance does not permit it.

How precise should we be when recording data? In other words, how many significant figures should we use? Rohlf and Sokal (230) recommended the following useful rule of thumb: *The number of unit steps from the smallest to the largest measurements should be between 30 and 300.* This means that when you deal with data having a small range, the use of more significant figures will reduce the percentage measurement error. When the range is greater, however, fewer significant figures can be used.

For example, if we have items that range from 6 to 9 centimeters, there are three unit steps between the highest and lowest measurements. An error of 1 would therefore produce 33 percent error. If, on the other hand, we record our data to two significant figures, say from 6.2 to 9.4 centimeters, then, by using the second significant figure as the unit step, we now have 32 unit steps instead of 3. This reduces the error percentage to about 3 percent.

2.2 STATISTICAL SYMBOLS

Before going further, we need to briefly discuss the use of statistical symbols. Until recently, the symbols used in a given textbook often seemed to be based more or less on the whims of the individual author. Although to some extent this is still true, efforts to standardize the use of symbols have resulted in the now widely accepted practice of using Greek symbols for *population* values and Roman letters for *sample* values. For example, the *mean* height of the entire *population* of adult males in the United States is represented by μ (mu), while the *mean* height of a *sample* of, say, fifty males drawn from that population is given the symbol $\overline{X}$. Thus, the mean is represented by μ or $\overline{X}$, depending upon whether it is a population mean or a sample mean. In a similar way, the variation *within a population* is represented by σ^2 (sigma), while S^2 is the symbol for the degree of variation *within a sample* drawn from that population.

In Chapter 5 you will find a more detailed discussion of populations and samples and their roles in inferential statistics. At that point you will again be reminded of the distinctions that must be made between Greek and Roman symbols.

2.3 DESCRIBING A DISTRIBUTION

A series of measurements or counts is called a *distribution*. Table 2.1 shows such a distribution, purposely kept very simple so that we can show some major concepts without getting all tangled up in arithmetic at this stage of the game.

First, note that the distribution in Table 2.1 consists of 8 numbers. This is the n number of the distribution. Therefore, $n = 8$.

The numbers in a distribution are also called *variates*, and in Table 2.1 they are denoted by upper-case X. The subscript (i) placed on the X in the first column heading is simply a general symbol that refers to the position

TABLE 2.1

X_i	$(X_i - \overline{X})$	$(X_i - \overline{X})^2$		
26 (X_1)	+ 19	361		
8 (X_2)	+ 1	1	$n =$	8
6 (X_3)	− 1	1	$\sum X_i =$	56
5 (X_4)	− 2	4	$\overline{X} =$	7
4 (X_5)	− 3	9	$\sum(X_i - \overline{X}) =$	0
3 (X_6)	− 4	16	$\sum(X_i - \overline{X})^2 =$	442
2 (X_7)	− 5	25	$S^2 =$	55.3
2 (X_8)	− 5	25	$S =$	7.4

of any given number or variate in the distribution. For example, look at the number 3, and you will see that $i = 6$ because 3 is the sixth variate down from the top.

If we add all the X_i's in the distribution, we obtain 56 as the sum. In many statistics books you will see this expressed with formal summation notation as follows:

$$\sum_{i=1}^{n} X_i = 56,$$

which you might think of as a computer program for that supercomputer you carry around in your head and which we hope you have with you now. The Greek letter sigma ($\sum$) is the summation sign, and it is one of the most frequently used symbols in statistics. The above instructions therefore tell you to *add* the numbers in the distribution, beginning with the first number ($i = 1$) and summing all of the numbers through n, which is another way of telling you to add *all* of the numbers in the distribution. On the other hand, the expression

$$\sum_{i=2}^{5} X_i$$

tells you to sum those numbers of the distribution beginning with X_2 and including X_5. In other words,

$$\sum_{i=2}^{5} X_i = 8 + 6 + 5 + 4.$$

Formal summation notation is a useful tool because it enables the statistician to express formulas and ideas in a general way without having to worry about specific examples. On the other hand, formal notation can become so complex that the biologist may spend more time wading through the symbols than he or she spends in learning the important concepts. We will therefore avoid the use of formal notation by assuming that in every case $\sum X$ implies $X_1 + X_2 + X_3 + \cdots + X_n$. Later, when things get slightly more complicated, we will use subscripts that will be kept as simple and descriptive as possible.

Now, let's turn back to our distribution in Table 2.1. The sum of the X's is 56 and dividing $\sum X$ by n yields 7. This value is called the *mean* and is symbolized by $\overline{X}$. The mean, which to you is an already familiar concept, is defined as *the value obtained by summing all of the variates in a distribution and then dividing by the number of variates in that distribution.* Symbolically, the mean is represented by

$$\overline{X} = \frac{\sum_{i=1}^{n} X_i}{n},$$

which is a simple formula obtained from the definition.

Although the mean is commonly called the "average," it is actually only one kind of average, since it is only one *measure of central tendency* of a distribution. The *median* is also an "average," and it may be defined as that measurement which has 50 percent of the members of the distribution above it and 50 percent below it. Looking now at Table 2.1, it may be seen that the midpoint of our distribution must lie between 4 and 5. The median is therefore 4.5. It may also be called the 50th percentile, since 50 percent of the variates in the distribution are below it.

It can be seen that the median (4.5) does not have the same value as the mean (7). The reason for this becomes obvious if we consider changing the first number from 26 to 26 *million.* You can see that this would have a marked effect on the mean but *the median would remain at 4.5!* Yet mean and median both qualify to be called the "average" of the distribution. Since the mean is affected by extreme measurements and the median remains unaffected, we therefore have to be careful how we interpret the term "average." If you see an "average" quoted, you should ask yourself whether what is being discussed is the mean or the median, who is saying it, and what he or she is trying to show.

The *mode* is still another possible measure of central tendency. In French, *la mode* means "the fashion," and that which is most fashionable is that which is seen more often! In the distribution in Table 2.1, the number that appears most often is 2. Although there are three possible measures of central tendency of a distribution, in our work we will concentrate on the mean because it plays the most useful role in experimental statistics. Later, we will see why this is so.

So far, we have described the distribution in Table 2.1 in terms of central tendency. This does not complete the description, however, for another very important value remains to be calculated. Suppose two large groups take the same test (in statistics, naturally) and upon evaluating the results we find that both group means are 50. Does this imply that both groups are the same, relative to this test, or at least very similar? We might be tempted to think so unless we look further at the two distributions of scores. Suppose that, in the first group, all the scores fall very close to 50, and in the second group the scores range all the way from 0 to 100, with a sprinkling of 10's, 20's, 80's, 90's, etc. Now we see that, despite the identical means, one of these groups is homogeneous with respect to the statistics test, while the other group is definitely heterogeneous. Therefore, even though the means are the same, the nature and makeup of the two groups are actually quite different.

The second important descriptive feature of a distribution is therefore the *degree of dispersion* of the measurements around the mean. This measure is called the *variance* and Table 2.1 contains the data necessary to calculate it.

If we take the mean, $\overline{X} = 7$, and determine the deviation of each member of the distribution from it, we will obtain a column of deviations from the mean.* Note that each deviation is signed according to whether the individual measurement is larger (+) or smaller (−) than the mean.

Adding these deviations and keeping the signs in mind, we obtain zero as their sum. It is an important fact that the sum of signed deviations from the mean of any distribution of measurement data will always equal zero. It is also true that the sum of the deviations from any number other than $\overline{X}$ will not equal zero.

Our next step in calculating the variance is to square each deviation, resulting in the next column. Note that the squaring procedure gets rid of the minus signs. Adding these squared deviations yields 442, which is the sum of squared deviations from the mean. This may be symbolized by $\sum(X_i - \overline{X})^2$, or simply by SS or ssq, both of which are abbreviations for "sum of squares." You probably have in mind a situation in which you square numbers and then add their squares. In statistics, however, when we speak of "sum of squares," we are talking about the *sum of squared deviations from the mean*. It is important to keep this in mind.

Also, to save wear and tear on pen and hand, the statistician often uses $\sum x^2$ to symbolize $\sum(X - \overline{X})^2$. (Note that we have dropped the i subscript, in accordance with our promise to avoid formal notation.)

Finally, we are ready to compute the variance of the distribution in Table 2.1. This is calculated by dividing the "sum of squares" (442) by n (8)

---

* The *deviations* are sometimes symbolized with a lower-case x (e.g., $x_1, x_2, \ldots, x_n$), whereas actual scores are denoted X (e.g. $X_1, X_2, \ldots X_n$). In this volume we shall, wherever possible use SS for "sum of squares," but remember that you compute it by finding the *individual deviations* first, squaring each one, and then adding all the squares.

to obtain 55.3 as the variance. Using S^2 as our symbol for variance, we can write the definition formula as:

$$S^2 = \frac{\sum x^2}{n} \quad \text{or} \quad S^2 = \frac{\sum(X - \overline{X})^2}{n} \quad \text{or} \quad S^2 = \frac{SS}{n},$$

since x and $(X - \overline{X})$ mean the same thing, i.e., a deviation from the mean.

The variance is therefore a measure of dispersion around the mean of a distribution, and it is calculated by dividing the sum of squared deviations from the mean by n.

Having computed the variance, we can now extract the *square root* of the variance to obtain S, another very important value associated with our distribution in Table 2.1. Since the variance obtained above is 55.3,

$$S = \sqrt{S^2} = \sqrt{55.3} = 7.4.$$

This value, 7.4 is called the *standard deviation** and is an extremely important basic tool of statistics. In the next chapter we will see an important relationship between the standard deviation and the normal distribution, or the normal curve.

2.4. MACHINE COMPUTATION

The method for computing variance described in the preceding section is simple enough, but if you review the number of steps involved you can see that calculating the variance of a large mass of data by this method would be tedious and time-consuming. For this reason, various shortcuts have been developed. With a good desk calculator, for example, the work involved in computing variance can be cut to a fraction of what would otherwise be required. In using a calculator, we need a different version of our formula for variance. This "machine formula" is

$$S^2 = \frac{\sum X^2 - (\sum X)^2/n}{n}$$

or its algebraic equivalent

$$S^2 = \frac{n\sum X^2 - (\sum X)^2}{n^2},$$

which is preferred by some because it involves only one division process.

---

* The standard deviation is sometimes denoted by S.D. but we shall use S (for the standard deviation of the sample used in an experiment). The standard deviation of a *population* (when it is known) is denoted by a lower-case sigma (σ).

One "run" through the data with a calculator will yield $\sum X^2$ and $\sum X$. These quantities, along with n, can be substituted in one of the above formulas, and S^2 can be calculated very quickly.

There is nothing mysterious about these computing formulas; they result from the expansion of the term $\sum(X - \overline{X})^2$ in the basic definition formula, followed by algebraic manipulations to produce $\sum X^2 - (\sum X)^2/n$, which is an algebraic identity to $\sum(X - \overline{X})^2$.

2.5 CODING

If a desk calculator is not available, our work can be simplified by a coding procedure.

Suppose that we have a distribution such as that in Table 2.2. It can be simplified by subtracting a constant from each number. In this case, we subtract 47 from each member, and this yields a new distribution labeled X_c. We find the mean of this new distribution by X_c/n, obtaining $\overline{X} = -2$. If we now add back the constant (47), we obtain 45, which is the mean of the original distribution. In performing this operation, we have made use of the fact that *subtracting a constant from each member of a distribution will yield a new distribution, the mean of which, plus the constant, will yield the mean of the original distribution.*

TABLE 2.2

X	X_c	x_c^2		
50	+3	+5	25	
49	+2	+4	16	$n = 7$
47	0	+2	4	$\sum X_c = -14$
47	0	+2	4	$\sum x_c = 0$
46	−1	+1	1	$\sum x_c^2 = 156$
40	−7	−5	25	$S^2 = 22.27$
36	−11	−9	81	$S = 4.72$

We can also calculate the variance, and therefore the standard deviation, by applying the procedure of Section 2.3 directly to the coded distribution X_c. Thus, subtracting the mean (-2) of the coded distribution from each member yields the deviations from the mean labeled x_c. Squaring these, we get the associated x_c^2 values, and summing this column yields $\sum x_c^2$. We now divide $\sum x_c^2$ by $n = 7$ and obtain 22.27 as the variance. It is important to note that we do *not* add back the constant this time since the *variance of a coded distribution will always be the same as the variance of the original distribution.*

1. Code by subtracting 47.

2. $\overline{X} = \dfrac{\sum X_c}{n} + \text{(subtracted constant)} = -2 + 47 = 45.$

3. $S^2 = \dfrac{\sum x_c^2}{n} = \dfrac{156}{7} = 22.27;$

$S = \sqrt{S^2} = \sqrt{22.27} = 4.72,$

from which it was derived. Remember, when the *mean* of a distribution is calculated by coding, the subtracted constant is restored, but when the *variance* is calculated, the constant is not restored.

The chief advantage of coding is that it permits us to work with small numbers. While not as convenient as a calculator, it is still a useful work-saving device.

We have seen that describing a distribution involves the calculation of two basic values. First, the mean is computed as the *measure of central tendency.* Second, to avoid the pitfall met by the statistician who drowned in a river having a mean depth of two feet, we determine the degree to which the measurements vary around the mean; this *degree of dispersion* is measured as the variance.

In the next chapter we shall apply these concepts to one of the most significant products of human thought—the normal curve.

PROBLEMS

2.1 In the following distribution, identify the mean, the median, and the mode (see Section 2.2):

$$3, \ 14, \ 20, \ 2, \ 18, \ 3, \ 13, \ 12, \ 25.$$

2.2 In the following distribution, compute the mean, the median, the variance, and the standard deviation (see Section 2.2):

$$6, \ 12, \ 11, \ 7, \ 8, \ 10, \ 6, \ 12, \ 9.$$

2.3 In the following distribution, compute the mean and the standard deviation (see Section 2.2):

$$1, \ 8, \ 10, \ 9, \ 14, \ 13, \ 14, \ 12, \ 8, \ 11.$$

2.4 In the following distribution, compute the mean, the median, the variance, and the standard deviation, using the method of coding (see Section 2.4):

$$63, \ 68, \ 62, \ 66, \ 68, \ 67, \ 63, \ 70, \ 69, \ 73, \ 68, \ 67.$$

2.5 Determine the mean, the variance, and the standard deviation of the following distribution, using the coding procedure (see Section 2.4):

$$97, \ 78, \ 105, \ 99, \ 101, \ 92, \ 108, \ 83, \ 112.$$

2.6 Find the mean, the variance, and the standard deviation of the following distribution by the machine method (see Section 2.3):

81, 75, 79, 57, 64, 69, 70, 69, 77, 79,

61, 83, 67, 63, 64, 72, 73, 80, 75, 64,

76, 74, 72, 73, 65, 74, 70, 71, 69, 69,

60, 60, 73, 59, 70, 69, 67, 66, 66, 62,

65, 70, 69, 68, 68, 68, 67, 76, 74, 70.

CHAPTER THREE

The Normal Distribution

3.1 NORMALLY DISTRIBUTED DATA

Figure 3.1 shows a *histogram* based on measurements obtained from 120 lima beans drawn one at a time from a half-bushel container. Each bean was measured to the nearest millimeter along its longest axis. The histogram is constructed by plotting length as the abscissa and frequency as the ordinate. The area of each bar therefore represents the frequency of cases found between the real upper and lower limits of the associated measurement. Thus, since 65 cases are associated with a length of 20 mm, this means that 65 cases have measurement values between 19.50 mm and 20.50 mm.

It is important to note that, even with this relatively small sample, the majority of cases tend to cluster around the central portion of the histogram. Also, as we move toward the left on the abscissa, toward shorter lengths, the frequency decreases, and there is a corresponding decrease in frequency as we move to the right toward larger measurements.

Now, suppose that we have available a very large population of field mice. Suppose further that we have built a better mousetrap and we are able to catch every mouse in the population and measure the length of its tail to the nearest millimeter. A histogram plotted from our data would probably look very much like the one shown in Fig. 3.2. Keep in mind that this is a theoretical histogram based on a *very* large population of mice and a very good mousetrap!

Again, we can see a clustering of cases around the center, with a gradual tapering off as we move toward more extreme measurements in both directions.

It is a notable and highly useful fact that many traits found in nature are distributed in this fashion. Carried to its ultimate, we obtain the curve superimposed on the "mouse tail" histogram in Fig. 3.2.

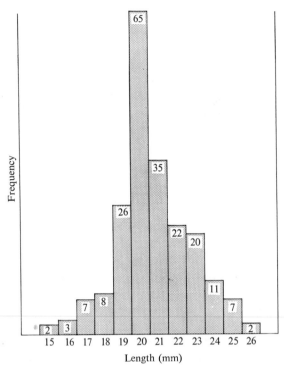

Fig. 3.1 Histogram based on measurements of 120 lima beans.

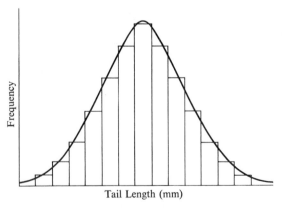

Fig. 3.2 Theoretical histogram based on mouse-tail measurements.

3.2. THE NORMAL CURVE

The curve shown in Fig. 3.3 is called the *normal curve*, and is without doubt one of the most useful theoretical tools ever discovered. The standard normal curve is bell-shaped, with the tails dipping down to the base line. In theory, they are asymptotic to the base line and do not touch it, proceeding *to infinity*. In practice, we usually ignore this, and work with practical limits.

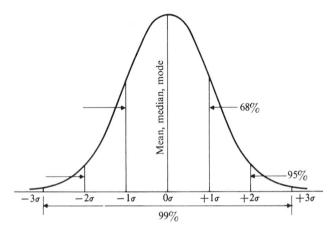

Fig. 3.3 Normal-curve relationship between standard deviation and area.

The perpendicular erected at the center of the curve in Fig. 3.3 represents the mean of the distribution. Since it obviously splits the distribution represented by the curve into *two equal parts*, it must also represent the median. Finally, we note that the frequency density is highest at this point, so the mode is also found at this same perpendicular. Thus, in the standard normal curve, the mean, median, and mode coincide.

3.3 NORMAL-CURVE AREAS

In Chapter 2 it was mentioned that the standard deviation bears an important relationship to the normal curve. Looking again at Fig. 3.3, we can see that the curve is divided into areas according to *standard deviations*.

Note that standard deviations to the right of the mean are (+), since they represent points *above* the mean; standard deviations to the left, or below the mean, are therefore (−). Figure 3.3 shows that approximately 68 percent of the total area of the curve is found between $+1\sigma$ and -1σ, 95 percent of the area is located between $+2\sigma$ and -2σ, and 99 percent of the curve is between $+3\sigma$ and -3σ. Note that we are using the Greek symbol (σ) for standard deviation because we are dealing with a *population*.

The question may arise as to *why* 68 percent of the curve area is contained between $+1$ and -1 standard deviations, and why 95 percent lies between $+2$ and -2 standard deviations, etc. This question may perhaps be best answered by pointing out that this is the mathematical nature of the curve, and may be compared with the Pythagorean theorem describing the relationship of the sides of a right triangle to its hypotenuse. We cannot "explain why" $c^2 = a^2 + b^2$, but it is nevertheless true and is a useful tool. The same may be said about the relationship of the standard deviation to the normal curve.

You will recall that any given area of the curve also represents a certain frequency of cases. Therefore, the entire curve area includes *all* cases in the distribution. That half of the curve to the right includes the upper 50 percent *of the cases*; 68 percent *of the cases* are included between $+1\sigma$ and -1σ, and so on.

To illustrate, suppose that we have a distribution with $N = 10,000$, $\mu = 50$, and $\sigma = 5$. What percent of the distribution would be below 55?

If $\sigma = 5$, then 55 would be one σ, or one standard deviation above the mean. We have just calculated a standard score, or z-score, and we used the following simple relationship:

$$Z = \frac{X - \mu}{\sigma} = \frac{55 - 50}{\sigma} = 1.$$

A standard score, or Z-score, is the *number of standard deviations* above or below the mean of a distribution.

It may be seen from Fig. 3.4 that since 55 is one standard deviation above the mean, approximately 34 percent of the area lies between 55 and the mean itself. Since 50 percent of the total area lies to the left of the mean, it follows that $50\% + 34\%$, or 84%, of the distribution lies below 55.

By similar reasoning, a measurement of 45 would be equivalent to a standard score of -1, or one standard deviation *below* the mean. Since 34 percent of the area lies between -1σ and the mean, 16 percent of the total distribution would lie below 45. Also, since $N = 10,000$, 16 percent of 10,000, or 1600, members of the distribution would lie below 45.

So far, so good. Now to get just a little more complicated, what percent of the curve area would lie below 57.50? Applying our formula for the standard score, we find that

$$Z = \frac{X - \mu}{\sigma} = \frac{57.50 - 50.00}{5} = \frac{7.50}{5} = 1.50.$$

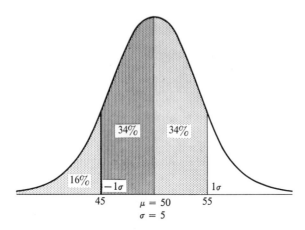

Fig. 3.4 Relationship of values of 55 and 45 to curve areas.

Since this is positive, 57.50 is situated 1.50 standard deviations above the mean. Now we must resist the temptation to interpolate between one and two standard deviations and their *associated areas*. Note that from the high point at the mean the curve dips sharply down toward the tails on either side. Thus, as we go in either direction from the mean along the base line, the *amount of area* contained between two perpendiculars will sharply decrease. Table IV in the appendix gives the area equivalent to various standard scores. Turning to Table IV, we find that a standard score of 1.50 is equivalent to an area of 0.4332, or 43.32 percent. Remember, the areas given in Table IV are always considered as located between the *mean* and a *perpendicular erected at the specific standard score involved.*

Thus if 57.50 is 1.5 standard deviations above the mean, and this implies that 43.32 percent of the curve area lies between 57.50 and the mean, it follows that 50% + 43.32%, or 93.32%, of the distribution lies below 57.70.

Looking at another kind of problem, what percent of the curve area lies between the standard scores +1.50 and +2.00? Again, we must resist the temptation to subtract 1.50 from 2.00 and look up the area equivalent to the resulting 0.50! Remember, the table areas are always between the perpendicular erected at the *mean* and the perpendicular erected at the standard score! Just to make certain you are convinced, let's look up 0.50. This turns out to be equivalent to 19.15 percent of the curve area, which is considerably greater than the actual area between +1.50 and +2.00 standard deviations, since between the two perpendiculars erected at these standard scores the curve dips sharply and the area between them is drastically reduced.

To find the correct area between +1.50 and +2.00 standard deviations, we first find the area between the mean and each of these standard scores. Thus, directly from Table IV, we find that

2.00   is equivalent to   47.72%,

and

1.50   is equivalent to   43.32%.

Subtracting 43.32 percent from 47.72 percent yields 4.40 percent, which is the correct area between +1.50 and +2.00 standard deviations. Figure 3.5 may help to clarify the foregoing example.

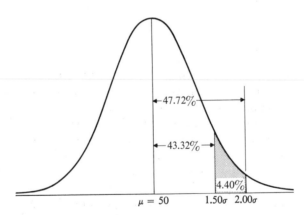

Fig. 3.5 Computing the area between 1.50 and 2.00 standard deviations.

3.4 DEPARTURES FROM NORMALITY

Table IV in the appendix, and thus our use of it, depends on a distribution having a shape similar to the standard normal curve. Not all distributions follow the standard curve to perfection, and in some cases, the departure from normality is enough to prevent our using the procedures described in this chapter.

Figure 3.6 illustrates a case of *skewness*; in this case, the curve is skewed to the left, or negatively skewed. Such a curve might be obtained from plotting age at death against frequency, since in the normal population there are many more deaths among older individuals than among younger ones. A curve skewed in the other direction would be skewed to the right,

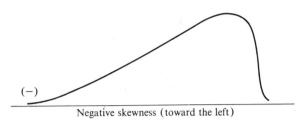

Negative skewness (toward the left)

Fig. 3.6 An illustration of skewness.

or positively skewed, having a preponderance of cases on the lower end of the scale.

Actually, it is rather amazing how far a distribution may depart from normality and still not affect statistical results to an appreciable degree. Badly skewed distributions can present problems; however, they can sometimes be "straightened out," or normalized, by special treatments of data called *transformations*.

The principles involved with the normal curve are basic to experimental statistics; applications of these principles will be considered in subsequent chapters. In the next chapter we shall consider probability and the application of normal-curve concepts to the calculation of probability.

PROBLEMS

3.1 From a large field of corn, 714 ears were collected in a random fashion. Each ear was measured to the nearest centimeter. Construct a histogram based on the data shown below (see Section 3.2):

Class interval	f	Class interval	f
24–25	11	16–17	220
22–23	45	14–15	100
20–21	80	12–13	60
18–19	190	10–11	8

3.2 Construct a histogram based on the following group data (see Section 3.2):

Class interval	f	Class interval	f
81–83	2	66–68	10
78–80	3	63–65	6
75–77	5	60–62	4
72–74	7	57–59	2
69–71	12		

3.3 Compute the Z-value of X in each of the following (see Section 3.4):

    a) $X = 40$, $\mu = 30$, $\sigma = 5$ b) $X = 51.50$, $\mu = 45$, $\sigma = 8$

    c) $X = 8.92$, $\mu = 10$, $\sigma = 3$ d) $X = 142$, $\mu = 159$, $\sigma = 12$

3.4 Compute the percentage of the normal-curve area that would be found *above* each of the following standard scores (see Section 3.4):

    a) 1.50   b) −1.50   c) 2.25   d) 1.9

3.5 Compute the percentage of the normal-curve area that would be found *below* each of the following standard scores (see Section 3.4):

    a) −0.40   b) 1.25   c) −2.00   d) 0.5

3.6 Compute the percentage of the area of the normal curve that would be found *between* each of the following standard scores (see Section 3.4):

    a) −0.50 and 0.50   b) 1.25 and 1.50

    c) −1.10 and −0.65   d) 1.40 and 1.65

3.7 Given the following standard scores, in each case compute the total area remaining in the tails of the normal curve (see Section 3.4):

    a) −1.64 and +1.64   b) −1.96 and +1.96

    c) −2.58 and +2.58   d) −0.95 and +1.25

CHAPTER FOUR

Probability

4.1 KINDS OF PROBABILITY

The term "probability" is one commonly used in ordinary conversation. In describing an event as probable, we imply that it is likely, or it is to be expected, or it is "in the cards." There is a difference, however, between stating, "It will probably rain tomorrow," and the statement, "The probability of rain tomorrow is 0.50."

In statistics, we attempt to express probability in precise quantitative terms. The basis for this quantitative expression may be built into the situation, as with a coin. In that event, it is called *a priori* probability, i.e., established "before the fact." Or we may formulate a quantitative statement based entirely on past experience; in this case we are dealing with *empirical* probability.

To illustrate, we assume, on the basis of meiotic division, that one-half the sperm produced by the human male carries the X-chromosome and one-half carries the Y-chromosome. If this is true, then the *a priori* probability that any random birth will produce a boy is $\frac{1}{2}$. On the basis of experience, however, we know that, over a period of time, more boys than girls are born, and the *empirical* probability that a boy will result from a random birth is actually about 0.51 in the white population of the United States. In fact, some studies of early abortuses indicate that the sex ratio (boys to girls) at conception may be as high as 160 to 100. This suggests that our *a priori* model is going wrong somewhere in the complex processes of reproduction.

A probability statement may predict all the way from certainty that an event *will* occur to certainty that it *will not* occur. If it is certain that a specific event will occur, the probability of its occurrence is 1. If there is no chance at all that it will occur, the probability is 0. Probability statements may therefore run from 0 to 1, and are usually expressed as fractions or decimals. It should be emphasized that while we may refer to "certainty" when tossing coins or throwing dice, in practical biological problems the concept of certainty is foreign to the careful experimenter. We are almost never certain of anything except the proverbial death and taxes!

4.2 INDEPENDENT EVENTS

Consider an honest coin, having heads on one side and tails on the other. If we assume that landing and balancing on edge is an impossibility, it follows that when such a coin is tossed it *must* land with either heads or tails facing up. Since there is no reason why one side should be favored over the other, the probability of obtaining a head is $\frac{1}{2}$ and the probability of a tail is also $\frac{1}{2}$. Since one *or* the other *must* occur, it follows logically that the probability of *either* a head *or* a tail coming up in any single toss of an honest coin is 1, or certainty.

Suppose that we toss a coin and it turns up heads. The probability of this happening is $\frac{1}{2}$. Now, if we toss the same coin again, what is the probability of a head on the second toss? A little reflection shows that, provided the coin is tossed in a random manner, the second toss will not be at all influenced by the results of the first toss. In other words, the second event is *independent* of the first event, and the two tosses are classified as independent events. It follows that if we toss an honest coin 10 times and obtain 10 heads in a row, although we might be inclined to bet our life's savings that a tail is "due" on the next toss, the probability of a tail on the eleventh toss is still $\frac{1}{2}$!

Suppose that we now toss a penny and a nickel together. What is the probability they will both turn up heads at the same time? Since the probability in each case is $\frac{1}{2}$, we have the following situation:

Penny	Nickel
H($\frac{1}{2}$)	H($\frac{1}{2}$)

Look at this situation from an intuitive point of view. While you might bet a considerable amount that a single coin would turn up heads, how much would you be willing to bet that a thousand coins tossed into the air would all turn up heads? You know instinctively that this is not a good bet, and extending the example to a million coins makes the point even more obvious. As the number of coins increases, the probability that they will all turn up heads decreases. Since probability is expressed in fractions, we will obviously multiply the separate probabilities of two or more independent events in order to obtain the *lesser* probability that they will occur simultaneously. In our example involving the nickel and penny we therefore *multiply* $\frac{1}{2}$ by $\frac{1}{2}$ to obtain $\frac{1}{4}$ as the probability that two heads will occur. This illustrates the *product rule* for calculating the probability that two or more independent events will occur simultaneously, or in a row.

What is the probability of obtaining a head and a tail when two coins are tossed? Consider the following:

Penny	Nickel
1. $H(\frac{1}{2})$	$T(\frac{1}{2})$
2. $T(\frac{1}{2})$	$H(\frac{1}{2})$

Note that in this case the desired event can be obtained in two ways, or by two different combinations. The penny can turn up heads and the nickel tails, or vice versa. Either combination satisfies the condition specifying one head and one tail. Calculating the separate probabilities of combinations (1) and (2), we have

$$\textbf{1.} \ \ \tfrac{1}{2} \times \tfrac{1}{2} = \tfrac{1}{4} \quad \text{and} \quad \textbf{2.} \ \ \tfrac{1}{2} \times \tfrac{1}{2} = \tfrac{1}{4}.$$

Since the event stipulated can happen as *either* (1) or (2), we *add* the probabilities of the ways by which a head-tail combination can be produced. Thus $\frac{1}{4} + \frac{1}{4} = \frac{1}{2}$, the probability of obtaining a head and a tail from a single toss of two coins or from tossing one coin twice in a row.

This is called the *addition rule,* and is applied to situations where a specific event can occur in more than one way.

It follows that the probability of two tails is the same as that for two heads, or $\frac{1}{4}$. Thus the probability of obtaining two heads, *or* one head and one tail, *or* two tails is $\frac{1}{4} + \frac{1}{2} + \frac{1}{4}$, or 1. In other words, if we toss two coins we are *certain* to obtain one of these three results.

Note how the product and addition rules are illustrated in the following example, involving three coins, where the possibilities are three heads, two heads and one tail, two tails and one head, or three tails:

$$
\begin{aligned}
&H(\tfrac{1}{2}) \times H(\tfrac{1}{2}) \times H(\tfrac{1}{2}) = \tfrac{1}{8} \\
&\left. \begin{aligned}
H(\tfrac{1}{2}) \times T(\tfrac{1}{2}) \times T(\tfrac{1}{2}) &= \tfrac{1}{8} \\
T(\tfrac{1}{2}) \times H(\tfrac{1}{2}) \times T(\tfrac{1}{2}) &= \tfrac{1}{8} \\
T(\tfrac{1}{2}) \times T(\tfrac{1}{2}) \times H(\tfrac{1}{2}) &= \tfrac{1}{8}
\end{aligned} \right\} \tfrac{3}{8} \\
&\left. \begin{aligned}
H(\tfrac{1}{2}) \times H(\tfrac{1}{2}) \times T(\tfrac{1}{2}) &= \tfrac{1}{8} \\
H(\tfrac{1}{2}) \times T(\tfrac{1}{2}) \times H(\tfrac{1}{2}) &= \tfrac{1}{8} \\
T(\tfrac{1}{2}) \times H(\tfrac{1}{2}) \times H(\tfrac{1}{2}) &= \tfrac{1}{8}
\end{aligned} \right\} \tfrac{3}{8} \\
&T(\tfrac{1}{2}) \times T(\tfrac{1}{2}) \times T(\tfrac{1}{2}) = \tfrac{1}{8} \\
&\overline{\text{Total} \qquad\qquad\qquad = 1}
\end{aligned}
$$

4.3 DEPENDENT EVENTS

In the situations seen thus far, a particular event has had no effect on the probabilities of subsequent events. Now, suppose that you choose a card from a well-shuffled deck and it happens to be an ace. The probability of drawing an ace was $\frac{4}{52}$, since there are four aces in a standard deck of 52 cards. Suppose, further, that you throw the ace out of the window and choose another card from the deck. Since the deck has been depleted by one card and one ace, the probability that the second card drawn is an ace is $\frac{3}{51}$. It may be seen that, in this case, the probability of the second event was indeed affected by the results of the first.

What is the probability of drawing three aces in a row from a deck of 52 cards *without replacement*? The probability that the first card is an ace is $\frac{4}{52}$, the second is $\frac{3}{51}$, and the third, $\frac{2}{50}$. The probability that these three events would occur in a row would therefore be found, as usual, by the product rule, or

$$\frac{4}{52} \times \frac{3}{51} \times \frac{2}{50} = \frac{3}{16,550}.$$

It may be seen that calculating probabilities involving dependent events follows the same rules as those involving independent events. The difference lies in the way that probabilities are assigned to the separate events.

4.4 APPLICATIONS TO GENETICS

The use of probabilities to predict the occurrence or recurrence of serious genetic defects is an important part of a relatively new branch of medicine known as *genetic counseling*. There are several reasons for the increase in importance of this specialty. First, it is a fact that genetically determined abnormalities constitute one of today's most significant health problems of childhood. In addition to the immeasurable emotional cost to the families of afflicted children, the general problem of birth defects places a considerable financial burden on affected families and on society in general. Secondly, it is now possible to *prenatally diagnose* certain genetic abnormalities and thereby provide the option of terminating the pregnancy at a relatively early stage. Third, potential parents who are carriers of certain recessive genes can now be identified by screening procedures. Advances in molecular genetics, cytogenetics, and biochemistry have therefore made it possible to apply probabilities to an increasing number of high-risk pregnancies and couples, and the genetics counselor is no longer restricted to scientific guessing games played *after* a defective child is born.

In this section we will discuss some simple but important examples of how the rules of probability can be applied to problems involving genetic abnormalities.

The Hardy–Weinberg Law

The Hardy–Weinberg law will be recalled by most biology students as a simple but extremely important principle that demonstrates that gene frequencies in a specific "ideal" population remain unchanged from one generation to the next. It may also be recalled that "violations" of the Hardy–Weinberg law, such as mutations, selection, small populations, etc., are part of the processes of evolution. Evolution, after all, is basically a phenomenon that is associated with changing gene frequencies.

To briefly review the Hardy–Weinberg law, let us consider a given pair of genes, A and a, where A is dominant to a. In a given population we can say that all of the A's and all of the a's will add to *all* of the A's and a's in that population, which is such a simple concept that it sounds like a silly statement. Therefore, if we let p stand for A and q stand for a, it follows that

$$p + q = 1$$

or, to put it differently, all of the p's and q's add up to 100 percent, or 1. Therefore, in a given population, the value of p is the frequency of gene A and the value of q is the frequency of gene a.

Going back to our genes A and a, we have the following genotypes, or gene combinations, in the population: AA, Aa, and aa. Using p^2 for AA, $2pq$ for Aa, and q^2 for aa, it then follows that

$$p^2 + 2pq + q^2 = 1$$
$$\text{(AA)} \quad \text{(Aa)} \quad \text{(aa)}$$

or, in other words, *all* of the AA's, Aa's, and aa's add up to *all* of them, which is 100 percent or 1. Thus, the Hardy–Weinberg law is a useful probability model; the value of p^2 is the probability that an individual randomly selected from the population is AA, and so on.

We can use this important concept to estimate carrier frequencies (Aa) in situations that involve recessive genes. For example, cystic fibrosis (CF) is a well-known genetic disease that affects children who have a "double dose" of the recessive gene (cc). CF, which is primarily a disease of Caucasoid groups, is a severe respiratory condition characterized by the accumulation of thick mucus in the respiratory tract. Since it often leads to death in childhood or early adolescence, we can assume that the vast majority of children with CF are produced by the following cross between two healthy carriers:

$$\text{Cc} \times \text{Cc}$$

yielding

$$\text{CC, Cc, Cc, cc.}$$

The frequency of CF is about one in 1600 live births. This, of course, is only an approximation based upon hospital records and is always subject to revision. Going back to the Hardy–Weinberg law, we can see that the frequency of homozygous recessive children (cc) is represented by q^2. Since q^2 is $\frac{1}{1600}$, q is therefore $\frac{1}{40}$. This is therefore the estimated frequency of the recessive gene for CF in the population. Now, if q is $\frac{1}{40}$, and $p + q = 1$, it follows that p is $\frac{39}{40}$. The value of $2pq$ is therefore very close to $\frac{1}{20}$. Since $2pq$ represents the carrier frequency, we can now assume that the probability that any randomly selected member of the population *is* a carrier (Cc) is $\frac{1}{20}$. Remember, however, that we should hedge a bit on this figure because it is based on the *reported* birth rate of CF, which is only a good (we hope!) approximation.

Now, let's take the case of Dick and Jane, neither of whom has any record of CF in their families. They are, therefore, "randomly selected" members of the population; thus, the probability that Dick and Jane are carriers is $\frac{1}{20}$ in each case. Therefore, the probability that *both* Dick and Jane are carriers is

$$\underset{\frac{1}{20}}{\underline{\text{Dick}}} \times \underset{\frac{1}{2}}{\underline{\text{Jane}}} = \frac{1}{400},$$

which you can see is similar to computing the probability that two coins will both come up heads at the same time.

Yet, even if they both *are* carriers and are therefore "eligible" to produce a CF child, the probability that they will do so is still "only" $\frac{1}{4}$, because the cross between two carriers produces the probability model

$$CC, \quad Cc, \quad Cc, \quad cc,$$

which tells us that with each independent birth the probability of a CF child (cc) is $\frac{1}{4}$, the probability of a normal child (CC or Cc) is $\frac{3}{4}$, and the probability of a carrier of cystic fibrosis (Cc) is $\frac{1}{2}$. Therefore, there are two events that must occur for Dick and Jane to have a CF child. First, they both must be carriers ($\frac{1}{400}$), and in any given conception the two recessive genes must get together ($\frac{1}{4}$). The product rule therefore again applies and $\frac{1}{400} \times \frac{1}{4} = \frac{1}{1600}$, which is the probability that any given child produced by Dick and Jane will have cystic fibrosis.

A more practical situation would be one in which Dick has a sister with CF. Having heard that CF is inherited, Dick and Jane are naturally concerned about their chances of having a CF child. In this case we can begin by assuming that Dick's parents are both carriers of CF (Cc). We can assume this because (1) they are normal, and (2) it is highly unlikely that *two* new mutations would have occurred at the same time. Since Dick is also normal he can not be cc. We can therefore eliminate cc from the genetic model, giving us

$$CC, \quad Cc, \quad Cc \quad \cancel{cc}.$$

Therefore, the probability that Dick is a carrier of CF (Cc) is now $\frac{2}{3}$, which is a considerable increase over $\frac{1}{20}$. If we now follow the same reasoning as before, we can see that the probability that Dick and Jane will produce a CF child with any given birth is $\frac{2}{3} \times \frac{1}{20} \times \frac{1}{4} = \frac{1}{120}$, which is a significant increase over $\frac{1}{1600}$.

Although marriage between first cousins tends to be frowned upon in many places by both church and statutory law, for the sake of illustration let us now assume that Dick and Jane are first cousins and thus have one set of grandparents in common (see Fig. 4.1). Now, if we take any given gene carried by their common grandmother, the probability that she passed this gene on to Jane's father is $\frac{1}{2}$. Then, if Jane's grandmother *did* give the gene to Jane's father, the probability that he in turn gave it to Jane is also $\frac{1}{2}$. It follows that for Jane to receive a given gene from her grandmother, two events must

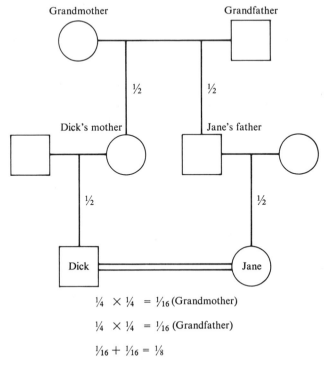

$$\frac{1}{4} \times \frac{1}{4} = \frac{1}{16} \text{ (Grandmother)}$$

$$\frac{1}{4} \times \frac{1}{4} = \frac{1}{16} \text{ (Grandfather)}$$

$$\frac{1}{16} + \frac{1}{16} = \frac{1}{8}$$

Fig. 4.1 Calculating the genetic relationship between first cousins.

occur in a row, giving us $\frac{1}{2} \times \frac{1}{2}$, or $\frac{1}{4}$, as the probability that Jane received a specified gene from her grandmother. Since Dick has the same grandmother, we can apply the same reasoning and obtain $\frac{1}{4}$ as the probability that Dick received a specified gene from his (and Jane's) grandmother.

What, then, is the probability that *both* Dick and Jane received a certain gene from their grandmother? Again, this is like finding the probability that two coins will come up heads at the same time. Therefore, $\frac{1}{4} \times \frac{1}{4} = \frac{1}{16}$, which is the probability that Dick and Jane have a specific gene in common. So far, however, we have not mentioned grandfather. Following similar reasoning, the probability that Dick and Jane have received a specified gene in common from their grandfather is also $\frac{1}{16}$. In other words, there are two possibilities whereby Dick and Jane could have a specified gene in common: (1) from grandmother $(\frac{1}{16})$, and (2) from grandfather $(\frac{1}{16})$. Adding the probabilities of the two possibilities we obtain $\frac{1}{16} + \frac{1}{16} = \frac{1}{8}$ as the probability that Dick and Jane have a given gene in common.

We therefore say that first cousins have $\frac{1}{8}$ of their genes in common. Putting it another way, if you are a carrier of a certain gene, then the probability that your first cousin is *also* a carrier is $\frac{1}{8}$. By applying the same reasoning illustrated in Fig. 4.1, you will find that brothers and sisters, mothers and sons, or fathers and daughters have $\frac{1}{2}$ of their genes in common, and that second cousins (the offspring of first cousins) have $\frac{1}{32}$ of their genes in common.

Now that we know that Dick and Jane are first cousins, the picture changes once again. Recall that the probability that Dick is a carrier of CF is $\frac{2}{3}$ (because of the existence of an afflicted sister). Because Jane is his first cousin, we can now assume that, given a specific gene carried by Dick, the probability that Jane also carries that gene is $\frac{1}{8}$. Therefore,

$$\frac{\text{Dick}}{\frac{2}{3}} \times \frac{\text{Jane}}{\frac{1}{8}} = \frac{1}{12}$$

and

$$\frac{1}{12} \times \frac{1}{4} = \frac{1}{48}$$

so the probability is now $\frac{1}{48}$ that any given child born to Dick and Jane will have cystic fibrosis.

The above example demonstrates that consanguineous matings (between relatives) may indeed increase the probability of defective offspring. This is certainly at least one explanation for the widespread sanctions against first-cousin marriages. It is also undoubtedly part of the reason for the practically universal ban on incest, even though the pharoahs of ancient Egypt often married their sisters because they were the only ones worthy to be their wives! Indeed, there are certain very low-frequency recessive genes that show up in the homozygous state almost exclusively as a result of consanguineous matings.

Tay–Sachs Disease

Tay–Sachs disease is a so-called metabolic-block disorder that involves a mutant gene that fails to code for an enzyme called hexoseaminidose A. Thus, the child that carries a "double dose" of this mutant gene fails to produce the enzyme, with the consequence that the nervous system progressively degenerates. Seemingly healthy at birth, the child develops the first sign of illness at approximately six months of age. Death comes inevitably in a slow, lingering fashion at about age four or five. Tay–Sachs disease is therefore an obviously devastating condition that places an incredible emotional and financial burden on the family. Like most serious genetic disorders, Tay–Sachs is not curable; the only hope lies in prevention of conception or termination of pregnancy if a positive diagnosis is obtained by amniocentesis.

This tragic disease provides an excellent example for our purposes here. First, it can be prenatally diagnosed. Secondly, carriers of the recessive Tay–Sachs gene can be identified by screening procedures. Finally, there is a reasonably identifiable population that is "at risk". The occurrence of Tay–Sachs disease is limited to a great extent to Ashkenazi Jews, i.e., Jews of European origin. For example, the birth rate of Tay–Sachs children among the Jewish population of the United States is estimated at $\frac{1}{2500}$. If we apply the Hardy–Weinberg law to this birth rate (q^2), as we did in the last section, we find that the carrier rate is approximately $\frac{1}{25}$.

The mating that produces a Tay–Sachs child is by necessity between two healthy carriers as follows:

$$Tt \times Tt,$$

which produces TT, Tt, Tt, and tt as the probability model which we can use when dealing with this condition. In this case, T is the normal gene and t is the mutant gene that fails to code for the appropriate enzyme. You can see that this probability model is identical to the one associated with cystic fibrosis.

For our case in point, let us assume that Henry and Ruth have been to a Tay–Sachs carrier-screening center and have been told that they are both carriers. Thus, the probability that Henry is a carrier is 1 and the probability that Ruth is a carrier is also 1. In other words, we must assume that the genotype of both Henry and Ruth is Tt. Referring to our probability model, we find that, with any given birth, the probability that Henry and Ruth will produce a Tay-Sachs child is $1 \times 1 \times \frac{1}{4}$, which is an extremely high probability if we consider the burden that Tay–Sachs disease implies. In other words, a probability of $\frac{1}{4}$ must be regarded very differently if we are talking about Tay–Sachs disease than if we were talking about the possibility of producing a child with large ears or an overbite. The latter two conditions can be corrected by plastic surgery or orthodontia.

Henry and Ruth are now faced with several possible alternatives. They can, of course, proceed to have children and take a chance that each time the genetic dice are rolled they will be winners. Suppose, for example, that they plan to have two children. What is the probability that at least one of the two will be afflicted with Tay–Sachs disease? Remember that *at least one* means *one or two*. If they have two children, Ellen and Renée, the probability that both Ellen and Renée will be afflicted is $\frac{1}{4} \times \frac{1}{4}$, or $\frac{1}{16}$. On the other hand, Ellen could be afflicted ($\frac{1}{4}$) and Renée not ($\frac{3}{4}$). The probability that this would happen is $\frac{1}{4} \times \frac{3}{4}$, or $\frac{3}{16}$. Finally, Renée could be afflicted ($\frac{1}{4}$) and Ellen not ($\frac{3}{4}$), again giving us a probability of $\frac{3}{16}$. Therefore, the total probability that *at least one* child will be afflicted is $\frac{1}{16} + \frac{3}{16} + \frac{3}{16}$, or $\frac{7}{16}$, which is coming very close to the flip of a coin. Using the same reasoning, satisfy yourself that the probabilities that at least one child in *three* will be afflicted, and that at least one in four will be afflicted, are $\frac{37}{64}$ and $\frac{175}{256}$, respectively. From the point of view of any competent gambler, it is apparent that Henry and Ruth are playing a losing game.

Of course, they may choose not to play the game at all by using effective contraception and adopting children if they wish to have a family. There is, however, still one more alternative. They may have each pregnancy monitored by amniocentesis and then choose to terminate the pregnancy if the fetus is diagnosed as a Tay–Sachs, or complete the pregnancy if it is not.

Huntington's Chorea

Huntington's Chorea is caused by a dominant gene. It is an especially insidious disease because its onset is usually delayed until middle age or later, *after* the victim has had children of his or her own. It is a lingering, degenerative illness that produces severe physical and mental disabilities before death finally releases its victims.

For our purposes, HC provides an interesting exercise in the applications of probability. Suppose, for example, that John Smith develops symptoms of HC at age 50. His daughter, Mary, is 28, and she has a five-year old son, Peter. Now that her father has Huntington's Chorea, what is the probability that Mary will also develop the disease when she reaches middle age? Since John's genotype is Hh, the probability that he gave Mary the dominant gene (H) for HC is $\frac{1}{2}$ because children receive one member of each pair of a parent's chromosomes. What about Peter? Again, two events must occur in a row: (1) John must have given the gene to Mary ($\frac{1}{2}$) and (2) Mary must in turn have given it to Peter ($\frac{1}{2}$). Therefore, the probability that Peter will some day come down with HC is $\frac{1}{2} \times \frac{1}{2} = \frac{1}{4}$.

Mary is therefore in the unenviable position of knowing that a flip of a coin will decide whether she will or will not develop HC. On the other hand, if she lives well past middle age without developing the symptoms, we can assume that she did not receive the gene from her father and Peter will then know that he, too, will live out his life free of the disease.

Huntington's Chorea is therefore an exercise in terrible uncertainty, considering the consequence of losing the coin toss. Woody Guthrie, a famous singer and composer of the early 1960's, died of HC at the age of 56. At the time of this writing, his children still await the verdict as to their own fates.

Hemophilia

Hemophilia is an X-linked condition that is characterized by excessive bleeding due to the lack of a factor in the clotting mechanism called the *antihemophilic factor* (AHF). The gene that controls the synthesis of AHF is located on the X-chromosome. There is also a mutant form of this gene that fails to properly code for AHF. Since the female normally has two X-chromosomes, in her case the mutant gene is recessive. For the male, who normally has one X-chromosome and one Y-chromosome, the presence of the mutant gene results in hemophilia.

Suppose that a woman with a hemophilic brother comes to us for an estimate of the probability that any son born to her will be hemophilic. Her parents are normal and she is married to a normal male.

We will start by assuming that the presence of the hemophilic brother implies that her mother is a carrier (Hh) of the recessive hemophilic gene. Since her father is normal, he is of the genotype YH. We may therefore assume that our counselee and her brother were produced by the following genetic cross:

$$YH \times Hh,$$

producing

$$YH, \quad Yh, \quad HH, \quad Hh,$$

where Yh is the hemophilic brother. The lady in question could be either HH or Hh. (Since she is a lady, she could hardly be YH or yh.) Therefore the probability that she is a carrier (Hh) must be $\frac{1}{2}$. She is married to a normal male (YH); therefore, if she is a carrier (Hh), we again obtain YH, Yh, HH, and Hh for the offspring. The probability that a male resulting from this cross would be hemophilic is therefore $\frac{1}{2}$, since we can discount the female genotypes.

Now it may be seen that two events must occur in a row. First, she must be a carrier ($p = \frac{1}{2}$). Second, if she is a carrier and produces a male, the probability that he is hemophilic is $\frac{1}{2}$. Applying the product rule, we obtain $\frac{1}{2} \times \frac{1}{2} = \frac{1}{4}$, which is the probability that a woman with her history will produce a hemophilic son. It should be noted that we ignored the possibilities involving female offspring since, in dealing with hemophilia, this would be of no concern, except for the possibility that a female could be a carrier.

This brings up another application of amniocentesis. A woman who knows that she is a carrier could have each pregnancy monitored, this time for the sex of the fetus. If the fetus is a female, there is no possibility that the child will be a hemophiliac (assuming that her husband is normal). On the other hand, if the fetus is a male, the probability that the child will have hemophilia is $\frac{1}{2}$, and a decision to terminate the pregnancy may then be made by the parents.

4.5 PERMUTATIONS

Suppose that George and John were each to take a certain course on a "pass–fail" basis. How many different possible combinations of grade could we expect? First, they both could pass. Secondly, John could pass and George could fail, or George could pass and John could fail. Finally, they could both fail. Thus,

George	John
P	P
P	F
F	P
F	F

gives us four possible combinations of grades. We could have found the same answer more quickly by using the "rule of counting." For example, in our case in point there are two possibilities for each individual. By multiplying these possibilities, we obtain the total number of possible combinations as follows:

John		George		
2	×	2	=	4.

As another example, how many triplets of RNA bases are available for use as codons in the process of protein synthesis? First of all, RNA contains the following bases: (1) adenine (A), (2) cytosine (C), (3) guanine (G), and (4) uracil (U). Each triplet has three positions, and four possible bases are available for position 1, four for position 2, and four for position 3. Now if we apply our rule of counting, we have:

Position 1		Position 2		Position 3	
4	×	4	×	4	= 64.

We therefore have 64 different triplet codons available, which are more than enough to handle the 20 or so different amino acids that are involved in protein synthesis.

Suppose, however, that we are interested in finding the total number of *orders* into which a certain number of objects can be arranged. For example, into how many different orders, or sequences, can the letters A, B, C, and D be arranged? To explain this, let's assume that we have four boxes labelled 1, 2, 3, and 4, as follows:

| 4 | | 3 | | 2 | | 1 |
|:-:| |:-:| |:-:| |:-:|
| 1 | | 2 | | 3 | | 4 |

Let us now further assume that we have the letters A, B, C, and D printed separately on cards so that we may put one letter in each of the boxes above. Starting with the first box, we can put in it any one of the four letters; i.e., at this point we have four possibilities. Now, having placed one of the letters in the first box, our possibilities for the second box are limited to the three letters remaining. When we reach the third box we have two possibilities remaining and, finally, we will have only one letter to drop into the fourth box. At this point we apply the rule of counting and multiply $4 \times 3 \times 2 \times 1$, which yields 24 different orders, or *permutations*, into which the letters can be arranged. If we had six letters, the solution would be $6 \times 5 \times 4 \times 3 \times 2 \times 1$, and so on.

This can be expressed in the form of a general formula for finding the number of orders, or *permutations*, into which n distinguishable objects can be arranged. This very simple formula is expressed as

$$n!$$

where the exclamation point (!) is a symbol for "factorial." To compute $n!$ we need only multiply together the numbers that begin with n and decrease to 1. Thus,

$2!$ is 2×1, $3!$ is $3 \times 2 \times 1$, $5!$ is $5 \times 4 \times 3 \times 2 \times 1$,   and so on.

Incidentally, $1!$ equals 1, and $0!$ is also considered to be 1.

To complicate things just a bit, how many permutations are there in the letters of the word BOOT? There are four letters altogether, but in this case the two O's constitute a class of indistinguishable objects. Quite obviously, it makes no difference whether we arrange two O's as OO or OO! We can see, therefore, that if we have one or more classes of indistinguishable objects, the total number of permutations will be reduced. This can be handled by the general formula

$$\frac{n!}{x!\,y!\,z! \cdots}$$

in which x, y, and z, etc., each represent the number of objects in a particular class of indistinguishable objects.

Consider, for example, our favorite word, STATISTICS. There are 10 letters altogether, but there are three S's, three T's and two I's. The total number of permutations is therefore

$$\frac{10!}{3!\,3!\,2!}$$

or

$$\frac{\overset{5}{\cancel{10}} \times \overset{3}{\cancel{9}} \times \overset{4}{\cancel{8}} \times 7 \times 6 \times 5 \times 4}{\cancel{3} \times \cancel{2} \times 1 \times \cancel{2} \times 1},$$

which yields 50,400 possible orders in which the word STATISTICS could be arranged even when three classes of indistinguishable objects are considered. Note that, when computing factorials, we can cancel between numerator and denominator to make calculations easier. If 50,400 different possible orders seem unbelievable, consider that a ten-letter word consisting of ten *different* letters has 3,628,800 permutations! If you are ever looking for a project for a rainy day, try listing the possible orders of the letters A through J!

Let's return to our boxes for a moment and suppose that we wish to find the number of orders in which A, B, C, D, and E can be arranged. This time, however, we will impose the conditions that C must always be first and D must always be last. This time, the C must go into the first box and the D must go into the last box. Therefore, after placing the C in the first box, we have only three possibilities available for the second box because we must save the D for the last box. Thus we have

which gives us 6 permutations, or possible orders when C must be first and D must be last.

To illustrate permutations with one more example, suppose that we use enzymes to break up a segment of an RNA molecule and find that it consists of 3 C's, 4 A's, 2 U's, and 3 G's. We will further assume that we also know that the segment begins with G and terminates with the fragment AAU. Of how many different orders of bases might this RNA segment consist?

First, adding up the bases, we find a total of twelve. However, one G is accounted for because we know that it is in the first position, and two A's and a U are accounted for because they make up the terminal fragment. This leaves us with eight bases to deal with; they consist of one U, 3 C's, 2 A's, and 2 G's. We therefore compute the total possible orders of bases as follows:

$$\frac{8!}{3!\,2!\,2!}.$$

Remember that the C's, A's, and G's represent classes of indistinguishable objects. Solving the above, we obtain 1680 possible base orders in the RNA segment.

4.6 THE BINOMIAL DISTRIBUTION

In Chapter 1 we described measurement data as continuous or discrete. Typical discrete data would be the number of heads in 500 tosses of a coin, the number of vestigial-winged fruit flies in an F_2 generation, or the number of children with sickle-cell anemia produced from 200 matings between carriers. These data are discrete, rather than continuous, because we cannot have 3.32 heads, 89.92 fruit flies, or 24.3804 children. Such data are distributed according to the *binomial distribution.*

To illustrate, let's return to our coin-tossing exercise. The probability that a coin will turn up either heads or tails is 1, or certainty. If p represents the probability of a head and q the probability of a tail, then $p + q = 1$. The probable distribution obtained from many tosses of a single coin can therefore be represented by $(p + q)^1$. The distribution resulting from tossing two coins would then be represented by $(p + q)^2$, and the distribution based on n tosses by $(p + q)^n$.

As an example, suppose that we toss five pennies a large number of times, recording the number of heads resulting from each toss. What distribution of heads could we expect? Expanding the expression $(p + q)^n$ or, in this case, $(p + q)^5$, we obtain:

$$p^5 + 5p^4q + 10p^3q^2 + 10p^2q^3 + 5pq^4 + q^5.$$

Keeping in mind that p represents the probability of obtaining a head in any one toss of a single coin, the probability of obtaining exactly five heads is represented by p^5, or $(\frac{1}{2})^5$, or $\frac{1}{32}$. The probability of four heads in any toss is derived from the expression $5p^4q$, or $5(\frac{1}{2})^4(\frac{1}{2})$, or $\frac{5}{32}$. Three heads is given by the expression $10p^3q^2$, or $10(\frac{1}{2})^3(\frac{1}{2})^2$, or $\frac{10}{32}$, and so on. The expression q^5 represents the probability of obtaining 0 heads, and at the same time represents the probability of obtaining five tails.

The result of the binomial expansion in the preceding example shows a definite symmetry, which is illustrated by the histogram in Fig. 4.2. Figure 4.3 shows a distribution of heads resulting from repeated tosses of ten pennies. Note that as n increases, the binomial distribution becomes smoother and approximates the normal curve.

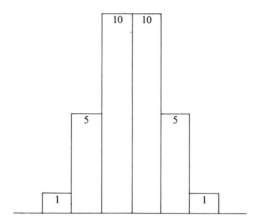

Fig. 4.2 Histogram based on expansion of binomial $(p+q)^5$.

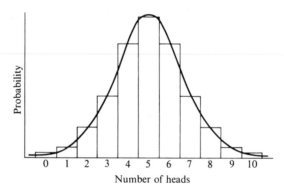

Fig. 4.3 Histogram based on repeated tossings of 10 pennies.

The degree of symmetry of the binomial distribution depends on the values of p and q. If p is $\frac{1}{2}$, and q is $\frac{1}{2}$, the resulting histogram and approximated curve will be symmetrical, as illustrated by Figs. 4.2 and 4.3. Figure 4.4 shows a histogram resulting from a distribution of p when p is $\frac{1}{4}$. This is a skewed distribution, which is not surprising since we would expect a greater density of q's. As p becomes less and q increases, the skewness also increases, so at very small values of p we encounter a different distribution, called the Poisson distribution, which is based on small values of p. This will be considered in detail in a later chapter.

It is not necessary to expand the binomial in order to calculate probabilities. Each coefficient found in the binomial expansion simply represents the number of possible combinations that can produce a given

Fig. 4.4 Skewed distribution when $p = \frac{1}{4}$, $q = \frac{3}{4}$.

event. These coefficients can be easily calculated using a simple mathematical trick from beginning algebra.

To illustrate, consider the probability of obtaining six heads in ten tosses of a coin. First, we can see that six heads and four tails are represented by $p^6 q^4$, where p and q are each $\frac{1}{2}$. But this event can occur through a number of different *combinations*, and it is necessary to calculate this number before we can proceed. Using the formula,

$$\frac{n!}{x! \, (n - x)!},$$

we can substitute 10! for $n!$, 6! for $x!$, and 4! for $(n - x)!$, or

$$\frac{10!}{6! \, 4!},$$

which is

$$\frac{10 \cdot 9 \cdot 8 \cdot 7 \cdot 6 \cdot 5 \cdot 4 \cdot 3 \cdot 2 \cdot 1}{(6 \cdot 5 \cdot 4 \cdot 3 \cdot 2 \cdot 1)(4 \cdot 3 \cdot 2 \cdot 1)}.$$

Cancelling and multiplying, we obtain 210 as the number of possible combinations that could produce six heads and four tails in ten tosses of a coin. Then $210p^6 q^4$, or $210(\frac{1}{2})^6(\frac{1}{2})^4$, yields $\frac{210}{1024}$, or 0.205.

4.7 CALCULATING PROBABILITY WITH A CURVE

It was fairly simple to calculate the probability of obtaining six heads in ten tosses of a coin by the factorial method, but suppose this problem were extended to 600 heads in 1000 tosses? The mechanics of such a calculation would be laborious, to say the least. In this section we shall consider a simple approach to calculating probabilities when n is large.

Looking again at Fig. 4.3, it may be seen that if ten coins are tossed a very large number of times, a very few tosses will yield 0 heads, and at the other end of the scale, an equally small frequency of 10 heads will be produced. The greatest frequency of tosses will cluster around the *mean* number of heads, or 5.

It may also be seen that even with an n as small as 10, the resulting binomial distribution is a rather good approximation of a normal curve. This means that we can use the normal curve as a basis for calculating the probability of obtaining six heads in ten tosses.

First, we calculate the mean of the binomial distribution by a formula based on the expression $(p + q)^n$,

$$\mu = np = \tfrac{1}{2} \times 10 = 5.$$

Second, we calculate the variance of the distribution by

$$\sigma^2 = npq = \tfrac{1}{2} \times \tfrac{1}{2} \times 10 = 2.50.$$

The standard deviation is found as usual by taking the square root of the variance; thus 2.50 yields 1.58 as the standard deviation.

Figure 4.5 shows the curve that is approximated by the binomial distribution of heads. Note that the area representing exactly 6 heads is bounded by the perpendiculars erected at 5.50 and 6.50, which are the lower and upper limits of 6. Knowing that the mean of the distribution is 5, and the standard deviation is 1.58, we can now use the procedures from Section 3.4 to calculate the area bounded by 5.5 and 6.5. Thus

$$Z = \frac{5.50 - 5.00}{1.58} = 0.32, \qquad Z = \frac{6.50 - 5.00}{1.58} = 0.95.$$

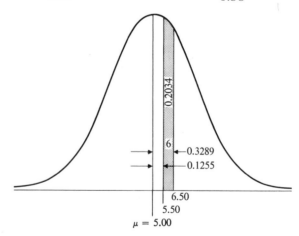

Fig. 4.5 The probability of obtaining six heads out of ten tosses of a coin.

From Table IV, 0.95 is equivalent to 0.3289, and 0.32 is equivalent to 0.1255. Since we want to know the shaded area in Fig. 4.5, we subtract 0.1255 from 0.3289 and obtain 0.2034. This is the probability of obtaining six heads in ten tosses of a coin. Comparing this with the value 0.205 obtained in Section 4.6, we can see that the difference is negligible even when working with a small n.

This method allows very simple and rapid calculation of the probabilities involving large n's, since the principle stays the same regardless of the number of coins or coin tosses.

Suppose that we wish to know the probability of obtaining *at least* six heads in ten tosses. This condition could be satisfied by 6, 7, 8, 9, *or* 10 heads. The probability of *at least* 6 heads would therefore be represented by all the area above the lower limit of 6, or 5.50, as shown in Fig. 4.5. Thus,

$$Z = \frac{5.5 - 5}{1.58} = 0.32.$$

Since 0.32 is equivalent to 0.1255 in terms of the normal-curve area, subtracting this from 0.5000 yields 0.3745 as the probability of obtaining at least 6 heads. As might be expected, this is a greater probability than that connected with *exactly* 6 heads, since the *possibilities* of obtaining at least 6 heads are greater. (See Fig. 4.6.)

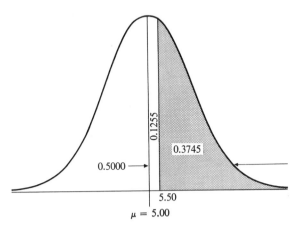

Fig. 4.6 The probability of obtaining *at least* six heads out of ten tosses of a coin.

What is the probability of tossing 1000 coins in the air and obtaining *exactly* 500 heads? At this point the uninitated is likely to shout "one-half!" since the probability of any single coin turning up heads is $\frac{1}{2}$. Looking at the

situation intuitively, how much would you be willing to bet that *exactly* 500 out of 1000 coins would turn up heads? Carrying this further, how would you like to bet your life's savings that *exactly* 500,000 coins out of a million will turn up heads? We can see intuitively that the probability of obtaining *exactly* the mean diminishes as n grows larger.

Calculating the probability of 500 heads in 1000 tosses, we have

$$\mu = np = \tfrac{1}{2} \times 1000 = 500,$$

and

$$\sigma = \sqrt{npq} = \sqrt{\tfrac{1}{2} \times \tfrac{1}{2} \times 1000} = \sqrt{250} = 15.80.$$

In Fig. 4.7, the area representing the probability of obtaining exactly the mean, or 500 heads, is the area bounded by 499.50 and 500.50. Again, calculating the standard scores, we have

$$Z = \frac{500.5 - 500}{15.8} = 0.032, \qquad Z = \frac{499.5 - 500}{15.8} = -0.032.$$

Since 0.032 is equivalent to an area of approximately 0.0120, we *add* 0.0120 and 0.0120, obtaining 0.0240 as the probability of obtaining exactly 500 heads in 1000 tosses. This is a far cry from 0.5000!

Taking yet another example, suppose that we cross normal-winged fruit flies that are carriers of the recessive gene for vestigial wings. Our cross is therefore

$$Vv \times Vv$$

$$F_2: \qquad VV, \quad Vv, \quad Vv, \quad vv.$$

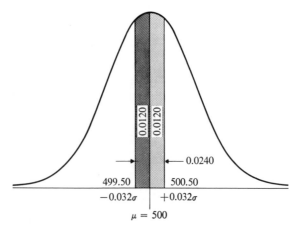

Fig. 4.7 The probability of obtaining 500 heads in 1000 tosses.

The probability that any single offspring has vestigial wings is therefore $\frac{1}{4}$, and the probability of its having normal wings is $\frac{3}{4}$. Since these probabilities differ from $\frac{1}{2}$, we could expect the distribution to be skewed. What effect would this skewness have on our calculations if we used the curve method to determine probability?

We will test this by calculating the probability of obtaining exactly 6 vestigial-winged flies out of a total of 12, assuming the genetic cross demonstrated above. By the more laborious factorial method,

$$\frac{12!}{6!\,6!}p^6 q^6,$$

where $p = \frac{1}{4}$ and $q = \frac{3}{4}$, we obtain 0.0400 as the required probability figure.

Using our skewed curve, we first calculate the mean,

$$\mu = np = \frac{1}{4} \times 12 = 3,$$

and the standard deviation,

$$\sigma = \sqrt{npq} = \sqrt{\tfrac{1}{4} \times \tfrac{3}{4} \times 12} = \sqrt{2.25} = 1.50.$$

Since the lower and upper limits of 6 are 5.5 and 6.5, we calculate the standard scores:

$$Z = \frac{5.5 - 3}{1.50} = 1.67, \qquad Z = \frac{6.5 - 3}{1.50} = 2.33.$$

Since 1.67 is equivalent to an area of 0.4525 and 2.33 is equivalent to 0.4901, we subtract and find the area representing 6 vestigial-winged flies to be 0.0376. This is sufficiently different from the previously calculated probability of 0.0400 to make us very cautious where skewed distributions are concerned. Actually, if n were larger, say around 40 or so, this discrepancy would be considerably reduced.

To summarize, if p and q differ from $\frac{1}{2}$, we should use the normal-curve approximation only if n is large. If p is extremely small, we must go to another distribution—the Poisson distribution—which will be considered in a later chapter.

In the next chapter we shall apply what we have learned about probability and distributions to a consideration of *inference*—a concept that is really the essence of experimental statistics.

PROBLEMS

4.1 A fair coin is tossed three times. Compute the probability of obtaining:

    a) A tail first, then two heads.

    b) Two heads and a tail, with order of occurrence unspecified (See Section 4.2.)

4.2 A die is rolled three times. Compute the probability of obtaining a 5 on the first roll, a 6 on the second roll, and anything but a 3 on the third roll (see Section 4.2).

4.3 A pair of dice is rolled twice. What is the probability of rolling either a 7 or an 11 on each roll? In other words, what is the probability of winning twice in a row? (See Section 4.2.)

4.4 Phenylketonuria is inherited as a simple autosomal recessive trait. What is the probability that two normal persons will produce a PKU child if we know that both sets of grandparents are carriers? (See Section 4.2.)

4.5 Assume that the couple in Problem 4.4 has three children. What is the probability that at least one of the three will be afflicted with PKU? (See Section 4.2.)

4.6 What is the probability of drawing three cards of the same suit from a deck, one at a time, without replacement? (See Section 4.3.)

4.7 Galactosemia is a genetic disorder that blocks the conversion of galactose to glucose. The condition is caused by a recessive gene, and occurs once in approximately 62,500 births. John Jones has a sister who is afflicted with galactosemia. John and both of his parents are normal. John and his wife, Mary, are not related, and there is no history of the condition in Mary's family. What is the probability that any given child born to John and Mary will be afflicted with galactosemia? (See Section 4.4.)

4.8 Assume that in Problem (4.7), John and Mary are second cousins. If this were the case, what would be the probability that any given child born to John and Mary will be afflicted with galactosemia? (See Section 4.4.)

4.9 Migraine headaches are apparently inherited through a dominant gene. Both Robert Smith and his mother are afflicted with migraine but his father is not. Robert's wife is also not afflicted. If Robert and his wife have two children, what is the probability that

    a) at least one of the two children will develop migraine?
    b) one or the other, but not both children will develop migraine? (See Section 4.4.)

4.10 A certain segment of an RNA molecule consists of 3 A's, 2 C's, 3 G's, and 2 U's. It is also known that the segment terminates with the fragment AG. The segment could be made up of how many different possible orders of bases? (See Section 4.5.)

4.11 From a box containing 3 white balls and 2 black balls, find the probability of drawing 2 black balls in a row if

    a) each ball is drawn with replacement;
    b) each ball is drawn without replacement. (See Sections 4.2 and 4.3.)

4.12 If a coin is tossed 8 times, what is the probability of obtaining at least 6 heads? (See Section 4.6.)

4.13 If a coin is tossed 6 times, what is the probability of obtaining at most 2 heads? (See Section 4.6.)

4.14 A certain disease has a mortality rate of 75%. Two patients suffering from the disease are selected at random. What is the probability that at least one of them will recover? (See Section 4.2.)

4.15 A dihybrid cross between AaBb and AaBb, where A and B are dominant, produces three offspring. What is the probability that at least two of the three will be of the genotype aabb? (See Section 4.2.)

4.16 A flower of genotype Bb is self-fertilized and produces 100 seeds. What is the probability that at least 60 of the seeds are of genotype bb? (See Section 4.7.)

4.17 Referring to Problem 4.16, what is the probability that, at most, 35 of the seeds will be of genotype bb? (See Section 4.7.)

CHAPTER FIVE

Inference

5.1 THE MEANING OF INFERENCE

The objective of most scientific investigations is to make general statements based on specific and relatively limited observations. A biologist interested in the effect of a certain drug on rat metabolism may try the drug on a few rats, using the data thus obtained to make an "educated guess" concerning the drug's effectiveness on rats in general. In doing so, he makes use of an important statistical concept known as *inference*.

Essentially, the methods of inferential statistics enable the experimenter to make limited statements concerning some characteristic of a *population*, based on data derived from only a part of that population. The term "population," as used here, refers to all possible observations that can be made on a specific characteristic. In our example above, this could mean all rats now living and all rats yet unborn. Or it could conceivably mean all rats of a certain species now living in a specific area, or simply all rats of a certain species. Essentially, the definition of what constitutes a population is up to the experimenter, and depends on the nature of the problem under investigation.

From a practical viewpoint, a population is usually a group so large that it precludes making direct observations. Instead, observations are made on a small segment of the population. This small segment is called a *sample*, and conclusions or estimates concerning the population are derived on the basis of the sample observations.

As an illustration, suppose that we wish to know the mean height of all adult males in the United States. We can now say that all adult males in the United States constitute our population. We can also see that to measure the height of each and every adult male in the United States is a practical impossibility, and the only way we will ever be able to make any statements concerning the mean of the population is through the use of inferential statistics.

5.2. SAMPLING

Our first step in attacking the problem is to draw a sample from our population of adult males. Now, it should be obvious that if our sample is to yield valid data concerning the population, the sample itself must be a fairly accurate cross section of the population. In other words, it is essential that a sample be *representative* of the population from which it is drawn.

Suppose that we visit racetracks around the country and measure all the jockeys we can find. Suppose further that we use the statistics gathered from this sample to estimate the true mean height of all adult males in the United States. Obviously we would be in error, with the error definitely on the short side! Or suppose we draw our sample from the ranks of professional basketball players. Again, our estimate of the population value would be way off, this time in the upward direction.

The jockeys and the basketball players illustrate the chief ingredient of a statistician's nightmare, i.e., the *biased* sample. Obviously, a biased sample is *not* representative of the population from which it was drawn. Furthermore, it is apparent that if we are to make reasonably valid estimates of population values based on samples, every attempt must be made to use samples which are representative of the population of interest.

Actually, we may never be *certain* that a sample is unbiased. If we could be sure of this, it would mean that we would know the exact nature of the population and would have no problem in the first place. In practical terms, we can only increase the chances of obtaining an unbiased sample by the way in which we select the individuals or measurements that comprise it. Usually, this involves selecting the sample randomly. *A random sample may be defined as a sample drawn in such a way as to ensure that every member of the population has an equal chance of being included.* Note that in the preceding illustration only jockeys or basketball players could be included; obviously these samples were no more random than they were representative.

A word of caution is in order at this point. A random sample is not *necessarily* always representative. For example, although the probability of doing so is small, by chance alone we could still draw a sample composed of basketball players! It is important to note that the term "random" as applied to a sample refers only to how it was drawn, and does not guarantee how representative it is. The terms random and representative are too often considered to be synonymous, when, in fact, they are not, Random sampling procedures increase the chances that a sample will be unbiased, and one of their major purposes is to minimize possible bias on the part of the experimenter.

Most experimenters are naturally going to be happier if they "prove" their hypotheses than if they fail to do so. No one is likely to gain fame and fortune for developing a drug that doesn't work or a hypothesis that proves untenable! The traditional "objectivity" of the scientist is probably more myth than fact, and scientific procedures demand that the experimenter show evidence that he has selected his samples and designed his experiment in order to avoid error through conscious or subconscious bias.

As a further illustration of the importance of randomness, let us suppose that we have 100 mice that we wish to divide into two groups of 50 each, in order to test a drug for its effect on heart rate. We might be tempted to close our eyes, reach into the cages, and "randomly" select 50 mice which would be assigned to the experimental group. The second 50, or those left, would become the control group.

What is wrong with this procedure? The major objection involves the fact that our experimental group is composed of the first 50 mice we were able to catch! Apparently they were not as lively as their brothers in the control group. Could this have anything to do with metabolic rate, and therefore have a profound effect on an experiment involving heart rate?

How could we have avoided this basic error in selecting our samples? For one thing, we could have assigned a number to each of the 100 mice, put corresponding numbers on slips of paper, mixed them up in a hat, and randomly selected the first 50 numbers; we would then assign these mice to either the control or experimental group as determined by a flip of a coin, or we could make use of the random numbers table, as described in Chapter 1. All 100 mice would then have had an equal chance to be included in the sample drawn, and we would have reduced the possible bias introduced by the experimenter or by the condition of the animals themselves.

The preceding example represents a rather simple problem in sampling. Unfortunately, obtaining a satisfactory random sample is not always so simple, especially for the biologist in the field. How does one collect a random sample of fish from a lake? Are the fish that allow themselves to be caught somehow different from the rest of the population? Similar problems arise in collecting samples of other fauna, whether by trapping, shooting, or other means. Actually, in cases like these it is probably well to admit that complete randomness in sampling is an *ideal*, toward which the investigator must strive, but which he will probably never fully achieve. Collecting plants from a particular locality in a random fashion is also a problem, but it is easier than collecting animals randomly, since plants stand still! Although some of the methods of field collecting are discussed later in conjunction with examples of statistical tests, a detailed account of this very complex area is beyond the scope of this book. The interested student is urged to consult the references cited in the bibliography.

5.3 SAMPLE SIZE

How does the size of the sample affect the accuracy of a generalization? Obviously, the larger the sample, the closer we are to measuring the population itself. Therefore, we may make the general statement that large samples are more useful than small samples *if* the large sample is *unbiased.* We can see that no matter how large a sample of jockeys we might have, any generalization based on it regarding the mean height of all United States adult males would be in error.

Essentially, a small unbiased sample is infinitely more useful than a large biased one. Logically, the *single* most important characteristic of a sample is not its size, but the extent to which it accurately represents the population from which it is drawn. Also, it will later be apparent from the logic of mathematics that the time and expense involved in increasing sample size beyond certain limits may result in diminishing returns.

5.4 STATISTICS AND PARAMETERS

Values such as means and standard deviations gained from samples are called *statistics.* In other words, sample attributes are statistics, and we speak of a sample mean as a *sample statistic.* Population values, on the other hand, are called *parameters.* Thus, in statistical inference, we draw samples from populations, derive sample statistics, and use these sample statistics as a basis for estimating unknown population parameters.

You will recall from Chapter 3 that statisticians generally use Greek letters to designate parameters and Roman letters to designate statistics. Thus the population mean is usually μ (mu) and the variance (σ^2) and standard deviation (σ) are symbolized by sigma. The sample mean is given the symbol $\overline{X}$, and the sample variance and standard deviation are written as S^2 and S, respectively. Also, the population number is generally represented by N, while a sample number is assigned small n.

5.5 THE SAMPLING DISTRIBUTION

In order to understand the logical basis of inference, it is very important to grasp the concept of the sampling distribution. It is likely that no other single concept is as basic to the procedures of experimental statistics. Before investigating sampling distributions, however, it might be helpful to review a couple of items considered previously. Figure 5.1 shows a population distribution; note the symbols used to designate the mean and standard deviation. Recalling the procedures of inference discussed earlier, we can visualize a sample distribution as representing a sample of measurements drawn from a population of measurements which are

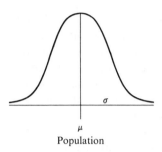

μ
Population

Fig. 5.1 A population distribution, showing appropriate symbols.

represented by the population distribution. Since the population is too large or is in some other way impossible to measure directly, we can assume that we do not know μ, and we probably do not know σ^2. Our problem, then, is to estimate on the basis of our sample statistics, $\overline{X}$ and S^2.

Before we become too involved in the actual estimation procedure, we should take a look at $\overline{X}$ and S^2 as estimators of population parameters. The sample mean, $\overline{X}$, is called an *unbiased estimate of* μ. This is because if we were to draw an infinite number of samples of a certain number n from the population, with replacement, *the mean of these sample means would equal* μ. (If the word "infinite" bothers you, a "billion billion" samples will probably do about as well.)

On the other hand, the mean of all the S^2's of an infinitude of samples would *not* equal σ^2. In fact, it would tend to be smaller than σ^2. For this reason, S^2, the sample variance, is called a *biased* estimate of σ^2. We are beginning to see that the word "biased," as used in statistics, means "you can't depend on it"!

Let us look at the biased nature of S^2 from a logical viewpoint. If we return to our overworked example of United States adult males, and assume a sample of 50 such males randomly drawn from the population, could we logically expect the degree of dispersion around $\overline{X}$ to be as great as it would be around μ? Obviously not; we would probably be surprised if our sample of 50 contained a man 4' 8" tall and another man 6' 8" tall! In other words, we are not as likely to find the various extremes of the population adequately represented in the sample.

If, therefore, we are going to use S^2 as an estimate of σ^2, it is apparent that we need to do something about making S^2 unbiased.

Remember that the basic formula for finding a sample variance is

$$S^2 = \frac{(X - \overline{X})^2}{n} = \frac{\text{SS}}{n}.$$

If 1 is taken away from n in the denominator, the value of S^2 would be increased, bringing it more in line with the population σ^2. Mathematical proof of this can be found in a book on mathematical statistics. It is actually rather simple, and is based on a principle found in Chapter 2; that is, the sum of the deviations from the true mean of a distribution will always equal zero, but the sum of the deviations from some number other than the mean will not equal zero. One does not have to be a mathematician, however, to see that this $(n - 1)$ factor is more important when using small samples; changing the denominator from 500 to 499 will hardly make an earth-shaking difference in S^2. In general, however, if we want to use statistics derived from a sample as estimates of population parameters, S^2 should be calculated as

$$S^2 = \frac{\sum(X - \overline{X})^2}{n - 1}.$$

Taking the square root of S^2 calculated in this manner will yield an estimate of σ that is more reliable.

Now, we return to our infinitude (or billion billion) samples drawn from our hypothetical population. Suppose that we were careful to record the mean of every sample drawn. This would yield a distribution of sample means which would take the form of a normal distribution. In other words, a few of these sample means would be extreme and would be found in the tails of the distribution. Most of the $\overline{X}$'s however, would tend to cluster around μ, the mean of the population from which the samples were drawn! This is sometimes called the central-limit theorem, and is one of the most basic and important principles in statistics.

It is highly unlikely that very many of our sample means would be *exactly* equal to the population mean, but we could probably expect that most of them would not deviate by too great an extent. A few, by chance alone, would be extreme deviates and be "way out" in the tails.

This distribution of *sample means* is called a *sampling distribution*, and is of critical importance to inferential statistics. Before someone thinks about all the work involved in drawing an infinite number of samples and decides to give up biometrics altogether, we should hasten to point out that the sampling distribution is a purely hypothetical concept, existing only in the minds of statisticians (and biologists interested in statistics). Like many such concepts in science, we shall shortly see that it is an extremely useful one.

It was pointed out previously that the mean of our sample means would equal μ. Therefore the mean of a sampling distribution and the mean of the population on which it is based are one and the same. Therefore μ may refer either to the mean of a population *or* to a distribution of means of samples drawn from that population.

Having identified the mean of our sampling distribution as μ, we now need to calculate the second important characteristic of any distribution— the variance. We can then take the square root of the variance to find the standard deviation. This will require some explaining, but if you understand this very important concept, you will later agree it was worth the effort.

5.6 THE STANDARD ERROR

Suppose that a sample consisting of one individual or measurement were drawn from a population. The mean of such a sample would be the same as the value of the measurement itself, and it could obviously vary around the mean of a sampling distribution of samples of $n = 1$ to the same extent it could vary around the mean of the population from which it was drawn. In other words, the variance of the sampling distribution in this case would be equal to the population variance, σ^2.

To illustrate, suppose a Martian is sent here to look over the situation preparatory to invading Earth. By chance, as he steps out of his flying saucer the first Earth man he happens to run into is a man 7′ 6″ tall! Immediately he rushes back to Mars and reports that Earth is populated by a race of giants! The invasion will likely be called off unless a little green man with a knowledge of statistics points out that a conclusion of this kind cannot be drawn on the basis of a sample of *one*.

On the other hand, suppose (and just suppose, for we are not likely to do it) that we draw a sample from a population that consists of *all* individuals or measurements in that population. Now, if we calulate our sample mean, it will obviously be equal to the population mean, and will not vary around μ at all. In other words, the variance of a sampling distribution based on the population N will be zero! This, of course, assumes another unlikely event; it assumes that all measurements were made with absolute accuracy.

From the foregoing, it is apparent that the extent to which a sample mean varies around the sampling-distribution mean will decrease as the sample size increases. We can also look at it in another way. If an infinitude of samples consists of large samples, we are not so likely to obtain extreme means as with samples that are smaller. Also, we note, from our case of the nervous Martian, that a small sample yields more room for error than a large one. If he had examined a reasonable sample of Earthlings, he would have been able to report on the size of Earth men in general with less chance for error.

Figure 5.2 shows how the shape of the sampling distribution becomes narrower as the sample size grows larger.

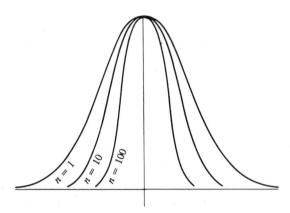

Fig. 5.2 Relationship between sample size and variance of sampling distribution.

Now we are ready to demonstrate a critically important basic formula. Recall that we said that the variance of samples of $n = 1$ would be the same as σ^2, or the variance of the population itself. If we let $\sigma_{\bar{X}}^2$ stand for the variance of the distribution of sample means, then

$$\sigma_{\bar{X}}^2 = \frac{\sigma^2}{1} \qquad \text{or} \qquad \sigma_{\bar{X}}^2 = \sigma^2.$$

On the other hand, we said that if sample size approaches the actual size of the population from which it is drawn, $\sigma_{\bar{X}}^2$ diminishes toward zero. Thus it becomes apparent that

$$\sigma_{\bar{X}}^2 = \frac{\sigma^2}{n}.$$

Taking the square root of both sides yields

$$\sigma_{\bar{X}} = \frac{\sigma}{\sqrt{n}},$$

which is the formula for the standard deviation of a sampling distribution of means. Since it is the standard deviation of a distribution of sample means, it is given the special designation *standard error of the mean.* Remember, however, that $\sigma_{\bar{X}}$ *is* a standard deviation, and its relationship to the normal distribution is the same as we learned in Chapter 3. It is simply a standard deviation of sample means in a sampling distribution, instead of in a distribution of individual measurements. (See Fig. 5.3.)

Now that we hopefully understand the sampling-distribution concept, what can we do with it? For one thing, we can now perform an important and useful kind of inference called *estimation.*

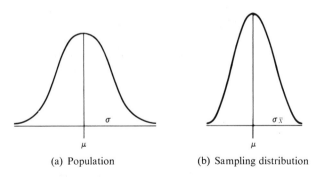

(a) Population   (b) Sampling distribution

Fig. 5.3 Population and sampling distributions.

5.7 ESTIMATION

Suppose that a sample is drawn from a population with an unknown μ and an unkown σ^2. Observations made on this sample yield the following statistics:

$$n = 100, \qquad \overline{X} = 50, \qquad S = 5.$$

Since we want to estimate μ on the basis of our sample statistics, we have derived S from an unbiased estimate of σ^2. In other words, S^2 was calculated as $\sum(X - \overline{X})^2/(n - 1)$.

Next, we will assume a hypothetical sampling distribution based upon an infinitude (or billion billion) samples of $n = 100$ drawn from the population of interest. We can now calculate the standard error of the mean, with the sample standard deviation subsituted for the unknown σ, and $S\overline{x}$ substituted for $\sigma\overline{x}$:

$$S\overline{x} = \frac{S}{\sqrt{n}} = \frac{5}{\sqrt{100}} = 0.50.$$

Once again, remember that the standard error (0.50) is also the *standard deviation* of the sampling distributions of means that would be obtained if we were to draw a billion billion samples of $n = 100$ from the population. Therefore, one $S\overline{x}$ to the right of the mean (μ) of the sampling distribution and one $S\overline{x}$ to the left of μ will encompass about 68 percent of the area under the curve. Or, to put it another way, plus and minus one standard error will include about 68 percent of the billion billion *sample means* that theoretically could be drawn from the population. It therefore follows that our sample mean ($\overline{X} = 50$) will be found between *plus and minus one $S\overline{x}$ 68 percent* of the time. Now, since $S\overline{x} = 0.50$, and *since our sample mean, $\overline{X}$, is an estimate of μ*, it follows that 68 percent of the time, μ will fall between 49.50 and 50.50. To put it another way, we can say with *68 per cent confidence* that μ falls between the upper limit 50.50 and the lower limit 49.50.

In scientific circles, however, this degree of confidence is not enough. Our estimation of μ will be looked upon more favorably if we can state a 95 or 99 percent confidence level. This change is simple to make. When we started our 68 per cent *confidence limits*, as they are called, we said that our sample mean ($\overline{X}$) would be found between ±1 standard error 68 percent of the time. In more symbolic form, we said that

$$\overline{X} \pm S\overline{x}(1) = 50 \pm 0.50(1);$$

therefore,

$$49.50 \leq \mu \leq 50.50.$$

If we want to change to 95 percent confidence limits, we must think in terms of 95 percent of the total area under the curve, or 47.50 percent on either side of μ. Looking at Table IV, it may be seen that an area of 47.50 percent is equivalent to 1.96 standard deviations (or standard errors). Therefore, to compute 95 percent confidence limits, we state that

$$\overline{X} \pm S\overline{x}(1.96) = 50 \pm 0.50(1.96)$$

$$= 50 \pm 0.98$$

or

$$49.02 \leq \mu \leq 50.98,$$

which represents the 95 percent limits of μ.

If we compare the 95 percent interval to the 68 percent confidence interval, it may be seen that the 95 percent confidence interval is larger. Logically, this makes sense because, to be *more confident* that μ falls within a certain interval, we must compensate by increasing the size of the interval between the lower and upper limits. In other words, the 95 percent confidence interval has less *precision* than the 68 percent confidence interval, assuming the same n-number.

Since $S\overline{x}$ equals S divided by $\sqrt{n}$, it may be seen that, as n increases, $S\overline{x}$ decreases. Looking back at the expression, $\overline{X} \pm S\overline{x}(1.96)$, it is apparent that, as $S\overline{x}$ decreases, the confidence interval must also decrease. Thus, as n increases, our estimate of μ becomes more precise. Again, this is logical because the larger the sample, the closer we come to the "truth," or to the actual population value.

It can therefore be seen that the precision of an estimate depends upon the level of confidence chosen and upon sample size. In other words, if we increase the level of confidence from, say, 95 percent to 99 percent, we must be satisfied with a wider interval which gives us a less precise estimate of μ. Since there is relatively little choice of confidence intervals, the more likely way to manipulate the degree of precision is to deal with sample size. By increasing the n-number, we can "zero in" on μ to a greater extent; even here, however, sample size is subject to practical considerations such as cost, time, and facilities. Furthermore, since the precision increases as the *square root* of *n*, an increase in *n* beyond a certain point will produce diminishing returns.

5.8 AN ESTIMATION EXAMPLE

Let us consider the following example as a summary of estimation procedures:

An investigator wishes to estimate the mean pulse rate of females between the ages of 18 and 21. A sample of 225 subjects is drawn randomly from the population consisting of females in the 18–21 age bracket. The sample statistics are computed, yielding the following:

$$\overline{X} = 76, \qquad S = 18, \qquad n = 225.$$

Now, we will estimate the population mean pulse rate (μ) step by step as follows:

1. First, we will compute the standard error of the mean as

$$S\bar{x} = \frac{S}{\sqrt{n}} = \frac{18}{15} = 1.20.$$

2. Second, we establish the upper and lower 95 percent confidence limits of μ by

$$\overline{X} \pm S\bar{x}(1.96),$$

$$76 \pm 1.20(1.96),$$

$$76 \pm 2.35.$$

3. The 95 percent confidence interval is therefore

$$73.65 \leq \mu \leq 78.35.$$

4. Suppose that we wished to estimate the mean pulse rate with 99 percent confidence instead of 95 percent? The procedure is exactly the same except that we will now have to deal with 99 percent of the curve, which includes 49.50 percent to the right of the mean and 49.50 percent to the left of the mean. Looking at Table IV, we can see that 49.50 percent of the curve to the right or left of the mean is equivalent to 2.58 standard deviations, or standard errors. The 99-percent confidence interval is therefore computed by

$$\overline{X} \pm S\bar{x}(2.58),$$

$$76 \pm 1.20(2.58),$$

$$76 \pm 3.10,$$

yielding

$$72.90 \leq \mu \leq 79.10,$$

which represents a wider interval, or a *less precise* estimate of μ. Again, this is logical because if we wish to be 99-percent confident instead of 95-percent confident, it follows that we must sacrifice some precision in our estimate. This assumes, of course, that we are dealing with the same n-number and the same sample standard deviation.

In the next chapter we will see how the ideas presented here can be extended to another statistical procedure that is extremely important to the experimental biologist—hypothesis testing.

PROBLEMS

5.1 Given the following distribution, calculate (a) the biased variance, and (b) the unbiased variance (see Section 5.5):

20, 19, 4, 8, 30, 24, 6, 18, 8, 19, 14, 10.

5.2 Given the following distribution, compute (a) the unbiased variance, and (b) the standard deviation (see Section 5.5):

142, 189, 172, 120, 198, 202, 150, 155,

168, 160, 145, 172, 162, 190, 150.

5.3 Compute the standard error of the mean of a sampling distribution based on a sample of 144, where the sample estimate of σ is 18 (see Section 5.6).

5.4 Compute the standard error of the mean of a sampling distribution based on a sample of 49, where the sample standard deviation is 14 (see Section 5.6).

5.5 Two hundred B.O.D. measurements, in milligrams/liter are taken at a specific location in a river. The mean B.O.D. is 95.50 and the sample S is 6. Estimate μ with 95% confidence (see Section 5.7).

5.6 Given the following sample data, compute

    a) the mean;
    b) the unbiased sample variance;
    c) the standard error of the mean;
    d) the 95% confidence limits of μ.

(See Sections 5.5, 5.6, and 5.7.)

72 75 77 79 86 89 92 40 65 54 55 42 52 63 40 36 62 96
48 51 49 54 55 59 64 95 46 62 48 51 55 78 38 92 46 66

5.7 A sample of 160 is drawn from a population. The sample mean is 60, and the unbiased sample variance is 25. Compute (a) the 95% confidence limits of μ, and (b) the 99% confidence limits of μ (see Section 5.7).

5.8 A sample of 100 is drawn from a population. The sample mean is 82.50, and the sample standard deviation is 12. Estimate μ with 99% confidence (see Section 5.7).

5.9 A sample of 225 is drawn from the same population referred to in Problem 5.8. Assume that the sample mean and the standard deviation are the same. Compute the 99% confidence limits of μ. How does the precision of your estimation compare with the precision of the estimation of μ in Problem 5.8? (See Section 5.7.)

5.10 A sample is drawn from a population where σ is known to be 13. What minimum sample size would have to be drawn in order to ensure a maximum range of 4.50 between the limits when estimating μ with 95% confidence? (See Section 5.7.)

5.11 A sample is drawn from a population where σ is known to be 18. What minimum sample size would have to be drawn in order to ensure a maximum range of 7.74 between the limits when estimating μ with 99% confidence? (See Section 5.7.)

CHAPTER SIX

Hypothesis Testing

6.1 DECISION MAKING

The experimental biologist is often required to make decisions or judgments concerning differences of various kinds. The taxonomist may wish to know whether certain morphological differences between populations are large enough to suggest subspeciation processes. The physiologist or clinician may be interested in the effectiveness of a specific drug on some variable such as heart rate or blood pressure. Further, it is usually necessary to make these judgments on the basis of samples drawn from populations that are too large to measure directly.

What constitutes a meaningful, or *significant* difference? Someone has defined a significant difference as "a difference that is large enough to make a difference"! This is not as silly as it may sound, since it is essentially the kind of judgment we have to make when dealing with certain kinds of experimental data.

It might be wise to point out a possible distinction between practical significance and statistical significance. As we shall see, statistical significance is determined in part by the experimenter and in part by the results of the statistical test. It is essentially arbitrary, and is mechanically determined. Practical significance, on the other hand, is determined by the biological knowledge and insight of the investigator as he interprets the statistical results in terms of the experimental setting. We should never forget that the statistical test is a tool, which the biologist must use judiciously. Statistical results will certainly help him make valid judgments, but they should never be allowed to blind him to various possibilities inherent in a specific experimental situation. Statistics is a useful slave, but it should never be permitted to become the master.

6.2 DECISION MAKING—AN EXAMPLE

Suppose that we toss a coin 100 times, and record the number of heads obtained. Suppose that we repeat this procedure for a very large number of trials. What is the expected mean number of heads?

Recalling the binomial distribution, the expected mean would be

$$\mu = np = 100 \times \tfrac{1}{2} = 50.$$

Figure 6.1 shows the normal-curve approximation resulting from many trials involving 100 tosses of an honest coin. Note that, in such a distribution, there is a possibility of zero heads and a possibility of 100 heads turning up, since these unlikely and unlooked for events *could* happen by chance.

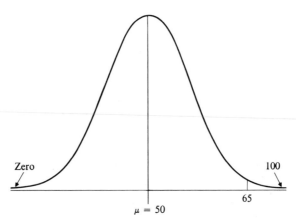

Fig. 6.1 Normal-curve approximation based on 100 tosses of an honest coin.

Now, suppose that, when we toss our coin, we obtain 65 heads. We are probably not too surprised, since we know that chance deviations from the mean would be the expected, rather than the unusual. Still, the more skeptical among us might wonder whether a difference this large could be due to a *causative factor*, such as a dishonest coin that is somehow biased in favor of heads!

Thus we have a problem. Is our coin a fair coin or is it biased in favor of heads? While we may never know the answer to this question, at least for certain, we can use statistical principles to help us arrive at an intelligent decision.

First of all, we must admit that it *is* possible to obtain even as many as 100 heads from 100 tosses of an honest coin *by chance alone*. We must therefore admit that 65 heads could possibly turn up without *necessarily* implying sinister machinations by a gambling syndicate. Still, you argue, the probability that 100 heads would turn up is no doubt so small as to be practically negligible, and you are willing to state *almost* without fear of contradiction that 100 heads out of 100 tosses would indicate a biased coin. Therefore, our problem really involves *finding the probability that an honest coin would produce 65 or more heads out of 100 tosses by chance alone.* Once this probability is determined, we can then make a decision about the fairness of the coin.

From Section 4.7, we are already familiar with this computation, and we can proceed to calculate the mean as 50, and the standard deviation as

$$\sigma = \sqrt{npq} = \sqrt{\tfrac{1}{2} \times \tfrac{1}{2} \times 100} = \sqrt{25.00} = 5.00.$$

Now, we place our value, 65 heads, on the distribution as illustrated in Fig. 6.2. Calculating the Z-value, using the lower limit of 65, we obtain

$$Z = \frac{X - \mu}{\sigma} = \frac{64.50 - 50.00}{5} = 2.90,$$

which is equivalent to a normal-curve area of 0.4981. The probability of obtaining 65 or more heads out of 100 tosses of an honest coin is therefore represented by the area *remaining* in the righthand tail: $0.5000 - 0.4981$ yields 0.0019 (see Fig. 6.2). Thus we find that the probability of obtaining 65 or more heads by chance alone is 0.0019, or $\frac{19}{10,000}$. Even the less conservative must admit that this is a pretty small probability, since we would expect to obtain 65 or more heads only about once in 500 trials.

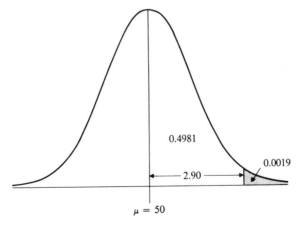

Fig. 6.2 Probability of 65 or more heads from 100 tosses of an honest coin.

What, therefore, is our most logical decision concerning the fairness of the coin? Since it has been established that the probability of an event "this bad or worse" occurring with an *honest* coin is only 0.0019, we may well be justified in deciding that our coin is indeed biased in the direction of heads.

Can we make this decision with complete confidence that we are correct? The answer is definitely *no*! Questions involving statistical decision making cannot be answered with certainty; the purpose of statistical hypothesis testing is to help us make intelligent judgments in the face of ever-present uncertainty. We have one distinct advantage, however; we can *estimate the degree of uncertainty.* In our example, the possibility of obtaining 65 or more heads with an honest coin *does* exist, and the probability of that possibility is 0.0019. Therefore, the probability that we have made an incorrect decision is 0.0019.

From the foregoing, it should be apparent that the word "prove" should be eliminated from the experimental biologist's vocabulary. He can never be *certain* his conclusions are correct, since he will always be haunted by a probability, however small, that an observed difference occurred by chance and not because of an assumed causative factor.

6.3 HYPOTHESIS TESTING

In the preceding section, we used statistical principles to reach a decision concerning the fairness of a coin. In doing so, we determined whether or not the difference between the number of heads *expected* and the number actually observed was statistically significant. In biological research, we often need to make decisions concerning differences, and we are helped to do this by an important procedure called *hypothesis testing*. In this section, we will go through the steps of a hypothesis-testing procedure, using as our example the rare situation in which we know the mean and standard deviation of a population. As you will see later, in most cases we do not know nor will we ever know the actual values of μ and σ. In cases that involve certain clinical values, however, we can sometimes assume at least a close approximation of μ and σ because of records of measurements obtained from millions of individuals.

Statement of the Problem

Suppose that a great many measurements taken over a long period of time lead us to assume that, in the general population of adult human males, the mean hemoglobin content is 15.80 g/100 ml of blood, with a standard deviation of 5. Now further suppose that we wish to gather information relative to the question of whether or not a prolonged exposure to lowered oxygen pressure at high altitudes produces a significant increase in blood hemoglobin level. We randomly draw a sample of 100 adult males from a population that lives in locations above 5000 feet. Tests performed on the sample yield a mean hemoglobin value of 15.96 g/100 ml. Our data may then be summarized as follows:

General population	Sample
μ_0 = 15.80 g/100 ml	n = 100
σ = 5 g/100 ml	$\bar{X}$ = 15.96 g/100 ml

The Null Hypothesis

In performing statistical tests of significance, the hypothesis is always stated in the null form. A *null hypothesis* is simply a statement of "no difference." Thus, in our hemoglobin problem, we should begin with the statement that "no statistically significant difference exists between the mean of the general *population* (μ_0) and the mean of the *population* living over 5000 feet (μ)," from which our sample was drawn. In terms of our specific question, we are saying that the mean hemoglobin content of adult males who live at high altitudes does not differ from that of the general population. To put it still another way, we are saying that living at high altitudes does not actually affect hemoglobin content, and that the difference that exists between the general population mean (μ_0) and the sample mean ($\bar{X}$) occurred by chance alone. Symbolically, we will express our *null hypothesis* as

$$H_0: \mu = \mu_0,$$

which says that there is no statistically significant difference between the general population mean (μ_0) and the mean of the special population (μ) from which our sample was drawn. Note that the symbols used in the statement of the null hypothesis are *parametric* symbols; that is, they refer to population, *not* sample values. In other words, we are not interested in the sample values for themselves alone; we are always interested in making "educated guesses" (inferences) about the populations themselves. To use an illustration, when the Salk vaccine for poliomyelitis was tested on a *sample* of children, the investigators were not interested in whether or not the vaccine effectively protected the children in the sample, but whether or not *all* children would be protected. This is the essence of inferential statistics.

In dealing with most biological research situations, we would like to show that a significant difference *does* exist. Naturally, no one wants to spend years developing a drug, only to demonstrate that it is not effective. A taxonomist who tries to show that two isolated populations of a certain species of bird are different to the point of demonstrating subspeciation would find few listeners if the two species turned out *not* to be different. The name of the game, therefore, is to be able to statistically *reject* the null hypothesis, thus obtaining evidence that a significant difference does exist.

The use of a null hypothesis is logically sound because one cannot "prove" a hypothesis. No matter how much evidence that supports a specific hypothesis is gathered, one cannot be certain that the same body of evidence would not equally support any number of unknown alternate hypotheses. On the other hand, it is logically possible to *reject* a hypothesis because this can be done if even one piece of evidence that contradicts it is found.

The Alternate Hypothesis

In our problem, the rejection of the null hypothesis, $H_0: \mu = \mu_0$, would allow us to state that our experimental and statistical results supports our contention that the alternate hypothesis, $H_A: \mu > \mu_0$, is true. Read that sentence again carefully and note the absence of the word "prove." We would, however, have experimental evidence to support the idea that adult males living at high altitudes actually tend to have a higher hemoglobin concentration than those found in the general population. Our conclusion must always assume, of course, that our experiment was carefully designed and carried out, that our data are valid, and that we were not wrong in rejecting the null hypothesis. Unfortunately, the possibility that we were wrong, no matter how small the probability of that possibility, will always be with us, like death and taxes. One must therefore beware of the experimenter who rejects a null hypothesis and then proclaims that he has thereby "proved" the alternate hypothesis.

Level of Significance

In our coin problem in Section 6.2, it was established that the probability of obtaining, by chance alone, a difference "as large as or larger than" the observed difference was 0.0019. We decided that this probability was too small to justify the conclusion that the observed difference was due to chance alone. Essentially, our decision at this point was to reject the null hypothesis that "no statistically significant difference exists between the observed and expected numbers of heads."

In formal hypothesis-testing procedures, we should establish, *prior* to performing the test, the *maximum* probability of a "chance alone" difference at which the null hypothesis will be rejected. In doing so, we are also stating the maximum acceptable probability that we will be wrong in rejecting the hypothesis.

The maximum probabilities, or *levels of significance*, that have been more or less arbitrarily established as acceptable rejection points are 0.05 and 0.01. In our hemoglobin-content problem we will set the level of significance at 0.05. Therefore,

$$\text{L.S.} = 0.05.$$

General principles involving levels of significance will be taken up in more detail in Section 6.5.

The Sampling Distribution

Our next step is to assume that this experiment has been performed an infinitude of times, yielding an infinitude of samples. From Chapter 5, we recall that the means of these samples would be normally distributed, and would constitute a sampling distribution of means as shown in Fig. 6.3.

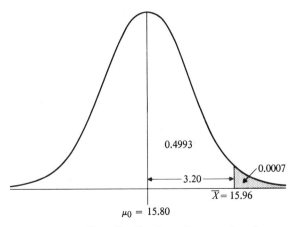

Fig. 6.3 Sampling distribution of means for the blood hemoglobin problem.

From the experimental data and from the null hypothesis, we assume that the mean of this sampling distribution is μ_0, or 15.80 g/100 ml. Our sample mean, $\overline{X}$, is 15.96 g/100 ml, and therefore falls on the distribution somewhere to the right of μ_0. Now we need to determine just how far to the right of μ_0, or how "far out," it is. In other words, we must place our statistic ($\overline{X}$) on the sampling distribution. This requires that we compute the standard error of our hypothetical sampling distribution. In computing the standard error, we can use σ in the formula developed in Section 5.6 because this is one of those unusual cases where we assume that we know the value of σ. Thus we have

$$\sigma_{\bar{X}} = \frac{\sigma}{\sqrt{n}} = \frac{5}{100} = 0.05.$$

Our next step is again a familiar one. To place our sample statistic ($\bar{X}$) on the sampling distribution, we calculate a Z-value in the usual manner (see Section 3.4). Thus

$$Z = \frac{\bar{X} - \mu_0}{\sigma_{\bar{X}}} = \frac{15.96 - 15.80}{0.05} = 3.20.$$

Having obtained 3.20 as the Z-value, we now know that our statistic ($\bar{X}$) lies 3.20 standard errors (or standard deviations) from the mean (μ_0) of our hypothetical sampling distribution of means. Turning now to Table IV, we can see that a Z-value of 3.20 is equivalent to 49.93 percent of the area of the curve. Looking at Fig. 6.3, we see that 0.0007 is therefore remaining in the righthand tail. Thus, the probability of obtaining a difference as large or larger than the observed difference between μ_0 and $\bar{X}$ by *chance alone* is 0.0007.

The Decision

We originally specified 0.05 as the maximum acceptable probability of being wrong if we rejected the null hypothesis. Since 0.0007 is considerably less than 0.05, our decision is to *reject* the null hypothesis H_0: $\mu = \mu_0$. You should also note at this point that an area of 0.05 in the righthand tail would leave 0.4500 between μ_0 and $\bar{X}$. This area (0.4500) is equivalent to a Z-value of 1.64. Therefore, if the Z-value turned out to be 1.64 or more, we would reject the null hypothesis (see Fig. 6.4). Since the Z-value that we actually obtained is 3.20, we are obviously justified in rejecting the null hypothesis.

Since 0.0007 is not only less than 0.05, but also less than 0.001, we can state our decision as follows: "The null hypothesis is rejected beyond the 0.001 level of significance." In practice, we generally speak of rejecting the null hypothesis at or beyond the 0.05 level, the 0.01 level, or the 0.001 level, whichever is dictated by the results of the test.

In our case in point, the probability of having rejected a *true* hypothesis is quite low; we can therefore feel reasonably safe in having made the decision to reject it. On the other hand, just so we do not become overconfident and given to statements that contain the word "prove," we must remind ourselves that, although the probability of having made the wrong decision is admittedly small, *it is not zero.*

In the case in point, we were able to reject the null hypothesis. If, on the other hand, we had obtained a Z-score of *less* than 1.64, an area *greater*

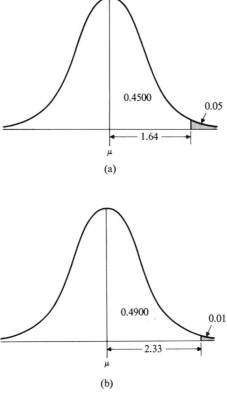

Fig. 6.4 Relationship between Z-values and the 0.05 and 0.01 rejection points.

than 0.05 would have been left in the tail. In that case, we would have *failed to reject the* null hypothesis. Note that we say "fail to reject" rather than "accept," since to accept a null hypothesis implies that no alternative hypotheses exist which would fit the situation equally well. This may be a bit of statistical pinhead polishing, and many respectable statistics references unashamedly use the word "accept."

Verbalizing Conclusions

In rejecting the null hypothesis H_0: $\mu = \mu_0$, we have demonstrated evidence that our sample ($\bar{X}$) was drawn from a population that is statistically significantly different from the general population, at least where hemoglobin concentration is concerned.

By demonstrating statistical significance, we have laid a foundation for arguments for *practical* significance. In practical terms, we are now prepared to suggest, *based on the data obtained from this specific experiment, and considering the limitations of this experiment,* that adult males who are exposed to prolonged periods of reduced oxygen pressure will develop a hemoglobin concentration that is higher than that found in the general population of adult males.

Note the use of "weasel words" in the preceding statement. The careful investigator hesitates, especially on the basis of limited data, to make sweeping statements. The phrase "beyond the shadow of a doubt," may be appropriate for the courtroom, but it has no place in scientific investigations where a shadow of a doubt *is always* there.

6.4 ONE- AND TWO-TAILED TESTS

In the decision problem described in Section 6.2, and again in the hypothesis-testing situation in Section 6.3, we were interested in whether our sample statistic was significantly *above* the mean of the distribution. In both cases, therefore, we were concerned with the area remaining in the right-hand tail only. This is called a one-tailed test and is appropriate if the *direction* of the difference is important or specified.

On the other hand, there are many situations where the investigator is concerned only with the significance of a difference, regardless of direction. If, for example, we were interested only in the question of whether μ differed significantly from μ_0, and were not concerned with its being larger (or smaller) than μ_0, then a two-tailed test would be appropriate. Figure 6.5(a) and (b) shows a two-tailed test situation. Note that since we will allow our statistic to go in either direction from the mean, the 0.05 probability of a "chance alone" difference is divided between the two tails, so we have 0.025 in the left tail and 0.025 in the right tail, comprising the total rejection region. Since this means that an area of 0.4750 or more must exist between the mean of the sampling distribution and the sample statistic, we note from Table IV that a minimum Z-value of 1.96 is necessary to reject the null hypothesis at the 0.05 level. Similar reasoning applied to the 0.01 level (0.005 in each tail) shows that a Z-value of 2.58 or more would be required for rejection with significance at or beyond the 0.01 level.

Since it requires a minimum Z-score of 1.96 to reject a two-tailed test at the 0.05 level, as opposed to the 1.64 required for rejection of a one-tailed test, it would appear to be more difficult to reject a null hypothesis when using a two-tailed test. Theoretically this is not really true, since, in performing a two-tailed test, we have two alternatives, or routes, to rejection. In the one-tailed test, on the other hand, we may reject only if our sample statistic falls in the area of rejection in one specific direction.

6.5 LEVELS OF SIGNIFICANCE

In Section 6.3 we described the level of significance as the probability of obtaining, *by chance alone,* a difference as great as or greater than the observed difference. We also said that the experimenter should present the maximum probability that can be accepted as justification for rejecting a null hypothesis.

Why did we select 0.05? Actually, there is nothing magical about it; in general, 0.05 and 0.01 are traditional levels that are considered compatible with sound experimental procedure. In a sense they are arbitrary, but they may also be said to have a basis in experience and logic.

Figure 6.5 illustrates the 0.05 and 0.01 levels of significance. Using normal-curve relationships, it may be seen that ±1.96 or more is the Z-value necessary to reject at the 0.05 level, using a two-tailed test. To

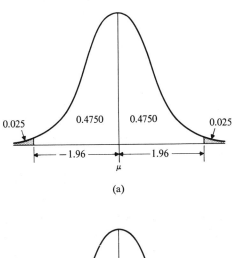

(a)

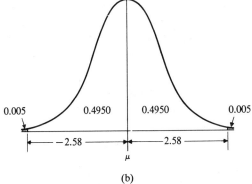

(b)

Fig. 6.5 Rejection areas and Z-values in two-tailed tests.

reject at the 0.01 level, however, a Z-value of at least ±2.58 is required, again assuming a two-tailed test. Thus it is more difficult to reject a null hypothesis at the 0.01 level than at the 0.05 level. Therefore, assuming that rejection of a specific null hypothesis is likely to lead the investigator to fame and fortune, it is obvious that the 0.01 level provides the more rigorous test and places the greatest "burden of proof" on the investigator. Setting a level of significance of 0.001 would quite obviously establish even greater rigor, but this is unsound because it would more likely lead to failure to reject null hypotheses when practical significance is actually present.

In the example described in Section 6.3, we rejected the null hypothesis "beyond the 0.001 level," actually at 0.0007. The probability that we were wrong in our decision to reject is therefore 0.0007, or "less than 0.001." If we were indeed wrong, then we were guilty of rejecting a hypothesis that was *true*, and we therefore committed a *Type I error*. The probability of a Type I error is specified by the results of the test; in our example it turned out to be "less than 0.001." Note that it is *not* 0.05, because 0.05 was simply the *maximum* probability of making such an error that we were willing to accept and that was specified before we began our analysis.

On the other hand, suppose that we had failed to reject the null hypothesis. Suppose further that we *should* have rejected it; i.e., the hypothesis was false. We then would have committed a *Type II* error. In other words, we would have failed to reject a false hypothesis.

Table 6.1 shows the relationship involving Type I and Type II errors. Note that the letters alpha (α) and beta (β) are used to denote the probabilities of Type I and Type II errors, respectively. As we have seen, alpha is specified by the level of significance obtained from the statistical test. The determination of beta is more complicated, but since failure to reject the null hypothesis usually implies an absence of practical significance, beta is not computed directly. "Success," in biological research situations, usually involves rejection of the null hypothesis; therefore, the investigator—and his critics—are usually more concerned about the probability of a Type I

TABLE 6.1

Type I and Type II Errors

	Reject	*Fail to reject*
Hypothesis true	Type I (α)	
Hypothesis false		Type II (β)

error, i.e., concluding that a drug is effective when it is not, etc. This is not to say that the investigator *always* wants to reject the null hypothesis. There are situations where so-called "negative" results may be important contributions to knowledge. As a generalization, however, rejection of the null hypothesis is the usual route to fame and fortune.

Figure 6.6(a) and (b) show a rather obvious relationship between Type I and Type II errors. As the probability of a Type I error increases, the probability of a Type II error decreases, and vice versa. Thus, one way to decrease beta is to increase alpha. This, of course, has its limitations, since the acceptable maximum at which we can establish alpha is 0.05. This does illustrate, however, that to insist upon a level of significance of 0.001 is to increase the probability of a Type II error to a needless extent. By setting 0.001 as the point of rejection, we would make too many Type II errors, and thereby miss too many "good things."

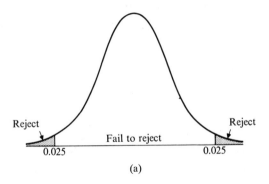

(a)

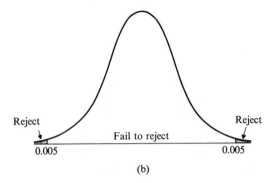

(b)

Fig. 6.6 Relationship between curve areas and probabilities of Type I and Type II errors.

A better way to decrease the probability of a Type II error is shown in Fig. 6.7. You will recall that the standard error of a distribution of sample means is found by

$$\sigma_{\bar{x}} = \frac{\sigma}{\sqrt{n}} \qquad \text{or} \qquad S_{\bar{x}} = \frac{S}{\sqrt{n}},$$

in which we use the sample standard deviation as an estimate of σ and in which n is the sample size. You should also recall that the standard error is actually the standard deviation of the distribution and is therefore a measure of dispersion of the sample means around μ. Looking at the above formula, we can see that, as n increases, $S_{\bar{x}}$ will decrease, producing a "narrower" distribution, as shown in Fig. 6.7(b). Note that the "fail to reject" region is smaller when n is larger and that the probability of a Type II error is thereby decreased. We could also decrease the value of $S_{\bar{x}}$ by decreasing S, but S represents an estimate of the natural variation in the population and is therefore a fact of life about which we can do very little.

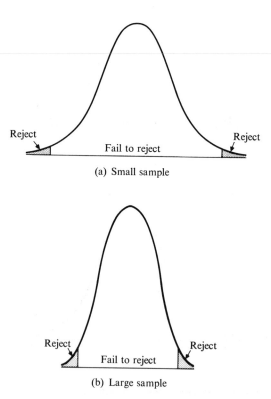

(a) Small sample

(b) Large sample

Fig. 6.7 Relationship between sample size and the probability of a Type II error.

Later on we will see that it is possible to reduce error from some sources, but we are always faced with a certain amount of error that is due to natural variation.

Statisticians often speak of the "power" of a test. We can define the power of a test in general terms as *the ability of the test to detect false null hypotheses*. Since *failure* to reject a null hypothesis is a Type II error, it follows that fewer Type II errors will be made as the power of a test increases. Again referring to Fig. 6.7, you can see that the test represented by Fig. 6.7(b) has more *power* than the test represented by Fig. 6.7(a). For a more extensive discussion of "power," the reader is referred to the text by Sokal and Rholf that is listed in the bibliography. From our own discussion, however, it may be seen that the power of a test is definitely associated with sample size.

On the other hand, remember that we are dealing with the *square root* of n; this means that there will be diminishing returns on our investment as we continue to increase sample size.

To illustrate this, let's assume that we wish to estimate the mean (μ) of a population with 95 percent confidence. To do so, we draw three samples with n-numbers of 100, 1000, and 10,000 respectively. We will assume that each of the three samples yields a mean of 50 and a standard deviation of 10.

Computing the 95 percent confidence intervals for each of the three samples, we obtain the following:

$$48.04 < \mu < 51.96 \quad (n = 100),$$

$$49.37 < \mu < 50.63 \quad (n = 1000),$$

$$49.80 < \mu < 50.20 \quad (n = 10,000).$$

Looking at the estimations based on sample sizes of 100, 1000, and 10,000, we can see that μ is estimated more precisely as sample size increases. We therefore have "better" estimations from larger samples. The question is, how much better? A little arithmetic will show that the confidence interval obtained from the sample of $n = 1000$ is 2.66 less than the interval based on the sample of $n = 100$. Looking further, however, we find that the interval decreased by only 0.86 as we went from $n = 1000$ to $n = 10,000$.

At this point, the researcher must ask an important question. Is the increase in precision between a sample of 1000 and a sample of 10,000 worth the very significant added cost in money, time, and effort? Rats, for example, are expensive and they eat a lot of expensive rat food. Also, the time required to carry out experimental procedures is vastly increased if we use a sample of 10,000 instead of 1000 or 100. What we have said about the precision of estimation can also be applied to the relationship between sample size and the power of a test. Increasing n does increase the ability of a test to detect false null hypotheses, but the problem of diminishing returns also applies. It is also true that, by using large enough samples, we could show that the difference between a population of adult males with a mean height of $5'\ 10''$ and one with a mean height of $5'\ 10\frac{1}{4}''$ is *statistically significant*. It is doubtful, however, that this difference would be regarded as *meaningful*, certainly not by bed manufacturers.

The reader is again referred to Sokal and Rholf for a more "scientific" way to determine minimum sample size, given certain conditions. It must be admitted, however, that, in practice, sample size is often determined by the demands and limitations of the situation. The story goes that someone once asked Abraham Lincoln how long a man's legs should be. Mr. Lincoln replied that they should be long enough to reach the ground. In a similar way, the size of a sample should be large enough to do the job and still consider the very practical limitations involving money and time that are imposed upon the researcher.

6.6 STUDENT'S t

Although the details may change with different problems and situations, hypothesis testing basically involves (1) locating a sample statistic on an appropriate sampling distribution, and (2) determining the relative distance of the statistic from the mean of that distribution. Statistical significance is then determined by interpreting the results of the general relationship,

$$\frac{\text{Sample statistic} - \text{Distribution mean}}{\text{Standard error}}.$$

In our example of Section 6.3, the above relationship produced a Z-value, since we were working with the normal-distribution values from Table IV.

At this point you should recall once again that the standard error of the mean is computed by

$$\sigma_{\bar{X}} = \frac{\sigma}{\sqrt{n}},$$

where σ is the population standard deviation. Since in most practical experimental situations we do *not* know the value of σ, we must use our sample standard deviation (S) as an estimate of σ.

Now we must consider that the value of the standard error determines where we place our sample statistic on the sampling distribution. The value of the standard error therefore determines how far out in "improbable land" our statistic falls, and this is the basis on which we decide to reject or fail to reject the null hypothesis. It follows that the validity of our decision depends to a meaningful extent on how much faith we can have in S as a good estimate of σ, since the value of S will significantly affect the value of the standard error.

If we randomly draw ten different samples, each with $n = 100$, from a given population, we will generally find the sample means to be very consistent and the sample variances to be also quite consistent. In other words, we would be likely to find only a relatively small variation among the ten sample variances. The sample standard deviation therefore appears to be a reliable estimate of σ when $n = 100$. Now, if we randomly draw ten different samples, each with $n = 10$, from the same given population, we will find the sample means to be still quite consistent but the variances will now be extremely inconsistent. For example, the sample variances might range from 80 to 650.

We must therefore conclude that a standard deviation obtained from a sample of $n = 10$ is *not* a reliable estimate of σ. Furthermore, if we use it in the formula

$$S_{\bar{X}} = \frac{S}{\sqrt{n}},$$

we will base our decision to reject the null hypothesis on a value of $S_{\bar{X}}$ which is much less reliable than it would have been if S had been obtained from a larger sample.

Unfortunately, the realities of biological research sometimes make it necessary either to work with small samples or to do nothing at all. Also, researchers often wish to use small samples to carry on preliminary, or "pilot" studies, which enable them to work the bugs out of experimental procedures and which also provide an idea of what the results from a more definitive experiment might be.

Fortunately for small-sample experimentation, a statistician by the name of W. S. Gosset, writing under the pseudonym "Student," developed a family of distributions which contain a "built-in" recognition of the limitations imposed by small samples. These "Student's t" distributions are associated with the concept of "degrees of freedom." Putting it simply, the degrees of freedom associated with a distribution are determined by the number of variates that can be entered in that distribution before the values of the remainder of the variates are fixed by the necessity to produce a certain total. This somewhat elusive concept may be clarified by considering the following distribution of five numbers:

$$\frac{X}{\begin{array}{c} 6 \\ 5 \\ 3 \\ 8 \\ 3 \end{array}}$$

It may be easily seen that the sum of this distribution is 25 and the mean is
5. Now, if we are to maintain these same sum and mean values, we may
enter four numbers of any value, but the value of the fifth number is fixed
by the necessity to produce a sum of 25 and a mean of 5. Since we have
"freedom to play around" with any *four* of these numbers, we say that this
distribution has four degrees of freedom.

When putting on a pair of gloves, we have a choice as to which glove,
right or left, we will put on first. Once the decision has been made,
however, we have no further choices; the other glove is "fixed." We might
therefore say that putting on gloves involves one degree of freedom!

A more basic discussion of degrees of freedom will be found in any
good book on mathematical statistics. For now, it will suffice to understand
that this concept is associated with sample size; i.e., the larger the sample,
the more degrees of freedom.

Figure 6.8 shows how the shapes of *t*-distributions are related to
sample size, or degrees of freedom. Note especially how the amount of
room in the tails increases as the sample size becomes smaller. Also note
that with increasing sample size, the curves increasingly approximate the
normal distribution of *Z*-values. This is logical, since, the larger the sample,
the more we are justified in using S as an estimate of σ.

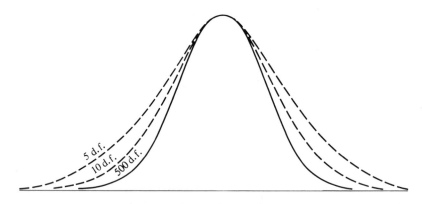

Fig. 6.8 Relation of shape of *t*-distribution to sample size.

Figure 6.9 shows that, as *n* decreases, we must work with curves having
proportionately more room in the tails. Note that, with a sample of 500, the
sample statistic must be located at least 1.96 standard errors from the mean
in order to reject H_0 at the 0.05 level. But, with a sample of 12, the statistic
must be located a minimum of 2.20 standard errors from the mean if we
are to be justified in rejecting H_0 at the same 0.05 level! A glance at Table
III in the appendix will show that the minimum, or *critical value* required

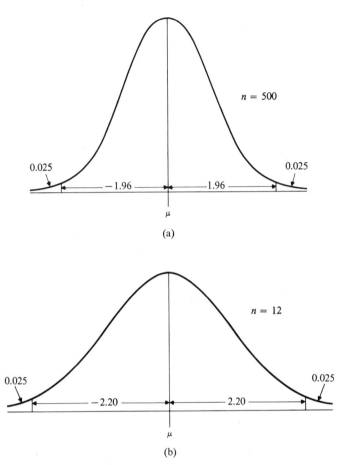

Fig. 6.9 Relation of sample size to the t-value necessary for rejection of H_0.

for rejection of the null hypothesis increases as degrees of freedom decrease.

The logic inherent in associating the t-curve's rejection values with sample size can be understood intuitively. If we are going to reject a null hypothesis on the basis of statistics gleaned from a small sample, then we need to use a larger critical value than would be necessary if our sample were large. In essence, the t-curves are telling us that small samples are not too reliable, and we must therefore go to greater lengths in order to justify rejecting a null hypothesis!

Referring again to Table III, we can see that, when using this table, we need to consider critical values of t that are specifically associated with appropriate degrees of freedom as well as a specific level of significance. Thus if we were working with a sample $n = 12$, and we wished to know the minimum, or critical value that must be attained for rejection at the 0.05 level, we would enter the table at $(n - 1)$, or 11 degrees of freedom. In the column labeled 0.05, we would find that the critical value is 2.20.

It is important to note that Table III has columns for both one- and two-tailed tests. Our value of 2.20 is the critical value for a two-tailed test at the 0.05 level. If a one-tailed test value is desired, then we must find the critical value in the next column to the left. In this case, look up the value associated with 11 degrees of freedom. This yields a one-tailed critical value of 1.80.

It is sometimes said that the t-distribution permits the effective use of small samples; i.e., as long as the t-test is used, small samples may be considered just as reliable as large samples. This is not really true! Large samples (within limits) are still better than small ones, and the major advantage of the t-distribution lies in the reduction of Type I errors that might otherwise result from the use of small samples. This, of course, is a useful advantage, but it still does not justify earth-shaking conclusions drawn from 7 or 8 specimens collected in a dubiously random manner! Perhaps at this point we need to again remind ourselves that statistical tests are no substitute for the common sense and biological knowledge of the investigator.

The following example illustrates how the t-distribution helps us to exercise caution when drawing conclusions based on small samples.

Example. A sample of $n = 225$ was drawn from a population and yielded an $\overline{X}$ of 50 and an S of 9. We wish to use this sample to estimate μ with 95 percent confidence.

1. Recalling the estimation procedures, we first compute the standard error of the mean by

$$S_{\overline{X}} = \frac{S}{\sqrt{n}} = \frac{9}{15} = 0.60.$$

2. Then μ is estimated by

$$\overline{X} \pm S_{\overline{X}}(1.96),$$

$$50 \pm 0.60(1.96),$$

$$50 \pm 1.18.$$

Therefore, $48.82 \leq \mu \leq 51.18$ is the confidence interval.

3. Now, instead of a sample of 225, suppose that we draw a sample, $n = 10$, from the same population and assume the same mean (50) and standard deviation (9).

4. Again, we find the standard error of the mean by

$$S_{\bar{X}} = \frac{S}{\sqrt{n}} = \frac{9}{3.16} = 2.85.$$

Note that the standard error associated with a sample of 10 is considerably larger than it is with a sample of 225.

5. We now find the so-called *critical value* of t that is associated with a two-tailed test at the 0.05 level. Entering Table III at $(n - 1)$, or nine degrees of freedom, we find this value to be 2.26.

6. Next, we follow the usual procedure for estimating μ, except that 95 percent of the t-curve associated with nine degrees of freedom encompasses ± 2.26 standard errors instead of ± 1.96 standard errors:

$$\bar{X} \pm S_{\bar{X}}(2.26),$$

$$50 \pm 3.16(226),$$

$$50 \pm 7.14.$$

Therefore, $42.86 \leq \mu \leq 57.14$ is the confidence interval.

From the foregoing example, it may be seen that, when we are working with small samples, the use of the t-distributions provides a realistic estimate of the precision (or the *lack* of it) that is associated with the use of small samples. It should be stressed, however, that this is *not* the same as saying that the t-table makes small samples "just as good" as larger samples that yield a better estimate of σ.

You should also note that the t-distribution table is "open on both ends"; that is, it may be used with large samples as well as with small samples. In fact, if you look at Table III, you will see that, as sample size increases, the critical values decrease until they become the same values that we have associated with the standard normal distribution. In practice, therefore, instead of computing areas from the normal-curve table, the experimenter simply consults the t-table directly, since, with given conditions, the critical value *is* the point at which the null hypothesis can be rejected.

6.7 ANALYSIS OF COMPLETELY RANDOMIZED TWO-GROUP DESIGNS

In Section 1.7 we developed the concept of simple two-group designs as part of our general discussion of the fundamentals of experimental design. You may recall that we discussed (1) the randomized design, in which different subjects were randomly assigned to each of two treatments, and

(2) the matched-pair design, in which subjects were matched in some way, possibly by using the same subject for both treatments. Before you continue with this and the next section, *it is important that you review Section 1.7.*

A common problem in biological research involves the determination of whether or not a statistically significant difference exists between the means of two populations. For example, suppose that two populations of a certain bird species appear to be reproductively isolated because of a geographic barrier. Suppose further that we suspect that independent evolutionary processes are going on in the two isolated populations and that two distinct varieties of the species now exist. For our example we will design a project that will test to see whether a significant difference exists between the mean tail lengths of the two isolated populations. In actual practice, a taxonomist probably would look at a combination of several morphological factors, but at this stage we will confine ourselves to one variable—tail length.

We now proceed to collect 24 specimens from population A and 21 specimens from population B, making every attempt to collect the specimens as randomly as possible. The tail length of each specimen is measured to the nearest one-hundredth of a millimeter, and we obtain the following sample data:

Population A	Population B
$X_A = 80.23$	$\overline{X}_B = 86.48$
$SS_A = 496.80$	$SS_B = 429.98$
$S_A^2 = 24.84$	$S_B^2 = 18.26$
$n_A = 21$	$n_B = 24$

Note that the figures 469.80 and 419.98 represent the sums of the squared deviations (SS) from each of the group means. Each of these sums of squares could be found very easily with a desk calculator by using the formula

$$\sum X^2 - \frac{(\sum X)^2}{n}.$$

The Null Hypothesis

Because we want to know whether a statistically significant difference exists between the *populations* from which the samples were drawn, the null hypothesis is stated as

$$H_0: \mu_A = \mu_B,$$

and, because it is hardly likely that we would go to all this trouble just to find that *no* difference exists between the population means, we hope to reject the null hypothesis and thus be able to say that we have experimental and statistical evidence to support the alternate hypothesis:

$$H_A: \mu_A \neq \mu_B,$$

which is a symbolic way of saying that the means of populations A and B *are* statistically significantly different. Once again, note the conspicuous absence of the word "prove" in the previous statement.

The Sampling Distribution

In Section 6.3 we were concerned with whether or not a sample mean was statistically significantly different from a population mean. We therefore tested the null hypothesis, $H_0: \mu = \mu_0$. In that situation our sample statistic was a mean, and we placed it on a sampling distribution of means. In our present case, however, we are concerned with a *difference between means*, so our sample statistic is $\overline{X}_A - \overline{X}_B$. We therefore construct a hypothetical sampling distribution of "differences between sample means." This is based on a hypothetical "infinitude" of samples drawn from the two bird populations. Also, since the null hypothesis implies *no* difference between the true population means μ_A and μ_B, the mean of this sampling distribution is assumed to be zero (Fig. 6.10).

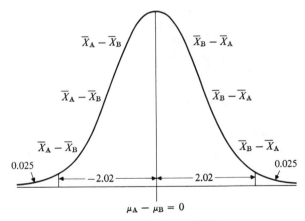

Fig. 6.10 Sampling distribution of differences between means.

Level of Significance

We shall set 0.05 as the maximum acceptable probability of making a Type I error. A two-tailed test is appropriate because we are interested only in whether or not there *is* a difference and not in the *direction* of the difference. Figure 6.10 therefore shows a rejection region of 0.025 in each tail; the total rejection region is 5 percent of the area under the curve.

Computations

1. Looking at the data derived from the samples drawn from the two bird populations, you will note that the sample variances are 24.84 and 18.26. Thus, while the two variances are not equal, they are relatively close, and an F-test (Chapter 8) would show that they are not statistically significantly different. Since the sample variance is an estimate of the population variance, in this situation we can assume that we are dealing with populations having similar variances. This is a basic assumption that must be met when making comparisons between or among populations. For example, suppose that we develop a new method of teaching biometrics and we decide to test its effectiveness by teaching class A with the new method and class B with a more traditional approach, thereby using class B as a control. At the end of a certain time period, we give the same examination to classes A and B and we discover that the test mean of class A is 60 and the test mean of class B is 50. Do we therefore conclude that our new method of teaching biometrics is superior to the traditional approach? Before we do, let us carry our example to an extreme and suppose that everyone in class A achieved a test grade of 60. Naturally, the mean of class A is therefore also 60. In class B, however, some students receive low grades, some are at the mean, and some are in the 80's and 90's. It is apparent from this that classes A and B are simply not comparable. In other words, classes A and B must have been drawn from very different populations in the first place. When we compare two populations with a t-test we must be able to assume that the variances of the two populations are similar. If they are not, then another test must be performed such as the Welch approximation described by Remington and Schork, found in the bibliography. In our case in point, the variances are not significantly different and we can therefore begin by *pooling* the two variances as follows:

$$S_p^2 = \frac{SS_A + SS_B}{n_A + n_B - 2}$$

$$= \frac{496.80 + 419.98}{21 + 24 - 2} = 21.32.$$

2. As usual, in order to place our sample statistic $(\overline{X}_A - \overline{X}_B)$ on the sampling distribution (Fig. 6.10), we must first compute the standard error of the statistic, which you should recall is also the standard deviation of the sampling distribution. Since we are working with a distribution of *differences* *between means*, the basic formula for the standard error is

$$\sigma_{\bar{X}_A - \bar{X}_B} = \sqrt{\frac{\sigma_A^2}{n_A} + \frac{\sigma_B^2}{n_B}}.$$

Of course, we do not know the values of σ_A^2 and σ_B^2, but if we can assume that they are similar, we can substitute the pooled variance (S_p^2) as an estimate of σ_A^2 and σ_B^2. Therefore,

$$S_{\bar{X}_A - \bar{X}_B} = \sqrt{\frac{S_p^2}{n_A} + \frac{S_p^2}{n_B}}$$

$$= \sqrt{\frac{21.32}{21} + \frac{21.32}{24}} = 1.38.$$

3. The number of degrees of freedom associated with this design is $n_A + n_B - 2$, which is the same as $(n_A - 1) + (n_B - 1)$. In our example we therefore have $21 + 24 - 2$, or 43 degrees of freedom. This brings up an important point: *As a general rule, we lose one degree of freedom for each parameter estimate used in our computations.* To illustrate, in our problem we have calculated the standard error as the square root of $(S_p^2/n_A) + (S_p^2/n_B)$. Now, we obtained S_p^2 by pooling S_A^2 and S_B^2, using the formula in step (1). When we calculated S_A^2, we used the mean of the sample $(\bar{X}_A)$, and we used $\bar{X}_B$ to calculate S_B^2. In other words we used two parameter estimates $(\bar{X}_A$ and $\bar{X}_B)$ to arrive at the value of the standard error in step (2). We therefore lose two degrees of freedom, one for each parameter estimate used in our computations. This is logical if you consider that it should be more difficult to reject a null hypothesis if we use (as we usually do) estimates of parameters instead of the true values of the parameters.

The Decision

We have calculated the standard error of the statistic as 1.38, and we can also see that the actual difference between our group means is $86.48 - 80.23$, or 6.25. We can now place our statistic $(\bar{X}_A - \bar{X}_B)$ on the sampling distribution by using the general formula

$$t = \frac{\text{Sample statistic} - \text{mean of sampling distribution}}{\text{Standard error of the statistic}},$$

which in this case translates to:

$$t = \frac{(\bar{X}_A - \bar{X}_B) - 0}{S_{\bar{X}_A - \bar{X}_B}}$$

$$= \frac{6.25}{1.38} = 4.53.$$

We now find the critical value of t associated with the 0.05-level two-tailed test with 43 degrees of freedom. Looking at Table III, we find that 43 degrees of freedom is not there. We therefore go to the next *lowest* degrees-of-freedom number, which turns out to be 40. Entering Table III at 40 degrees of freedom we find the critical value at the 0.05 level (two-tailed test) is 2.021. This critical value is the value that the t-value that was obtained above must *equal or exceed* if we are to reject the null hypothesis at the 0.05 level. Quite obviously, our value of 4.53 does exceed 2.021 and, in fact, it even exceeds the critical value (3.551) associated with the 0.001 level. We therefore reject the null hypothesis "beyond the 0.001 level"; this is a highly significant result because it means that we can reject the null hypothesis with a probability of less than 0.001 of having made a Type I error, that is, of having wrongly rejected the null hypothesis.

Considering this, we now feel confident as we claim that we have experimental and statistical evidence that supports our contention that the alternate hypothesis is true. In other words, we have experimental evidence that a significant difference in mean tail length exists between the two bird populations. Still once again, note the absence of the word "prove" in the statement above.

6.8 ANALYSIS OF TWO-GROUP DESIGNS—MATCHED PAIRS

Referring to Section 1.7, you should recall that certain extraneous independent variables may be better controlled by using the same subject for both treatments. Or, in some situations, subjects may be paired on the basis of genetic similarity, weight, age, or environmental and other factors.

Suppose that we wish to test whether or not epinephrine will significantly alter blood-cholesterol content in rats. We could, if necessary, randomly assign a portion of a sample of rats to an "epinephrine" group and the rest to a "placebo" group. As the procedure itself implies, this would be a completely randomized design and would be analyzed by the procedure described in Section 6.7.

Suppose, however, that we have good reason to assume that the effects of both the drug and the placebo are temporary; i.e., that neither treatment has a lasting effect on the blood-cholesterol level in the rat. We may therefore decide to give both treatments (at different times) to the same rat, using a number of different rats in order to satisfy the need for replications. We will also randomize the order of administration of drug and placebo to each rat by flipping a coin in each case. Finally, we arrange the data as shown in Table 6.2. We then proceed with the analysis as follows:

The Null Hypothesis

As usual, we test the hypothesis in its null, or "no difference" form. Since this is a matched-pair design, our statistic is the mean difference, $\bar{D}$, which is found by summing the differences between the placebo and drug responses for each replication and dividing by the number of *pairs*. Thus

$$\bar{D} = \frac{\Sigma D}{n}$$

Since we are stating that no significant difference exists between the drug and the placebo in terms of effect on cholesterol content, our null hypothesis in symbolic form is

$$H_0: \mu_{\bar{D}} = 0.$$

The alternate hypothesis that we hope to support by rejection of the null hypothesis is

$$H_0: \mu_{\bar{D}} \neq 0.$$

The Sampling Distribution

Figure 6.11 shows the sampling distribution associated with this test. This is a distribution of $\bar{D}$'s obtained from a hypothetical infinitude of experiments performed on samples of 12 drawn from the same population. Since we have stated $H_0: \mu_{\bar{D}} = 0$, the mean of this distribution is the mean of all sample $\bar{D}$'s, and is assumed to be zero. Since it is a mean of means, it may be symbolized by $\mu_{\bar{D}}$. We should also note that our sampling distribution is a *t*-curve based on 11 degrees of freedom, or 12 pairs minus 1.

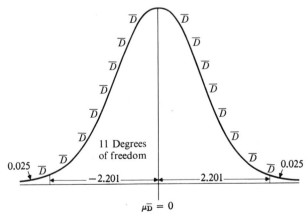

Fig. 6.11 Sampling distribution of $\bar{D}$ with 11 degrees of freedom.

Level of Significance

We will set the level of significance at 0.05. Since we are interested only in whether epinephrine will *change* cholesterol content, we shall assume a two-tailed test, as shown by Fig. 6.11.

Computations

1. Our first step is to find the difference D between the measurements taken on each replication. Note that this is a signed difference, and is (+) if the drug response is larger, and (−) if the placebo response is larger. Actually, it could be done the other way around, but a casual examination of the data listed in Table 6.2 indicates a trend toward larger values as we go from placebo to drug, so this way we simply avoid working with minus signs.

TABLE 6.2

Subject	Placebo	Epinephrine	D	$(D - \bar{D})$	$(D - \bar{D})^2$
1	178	184	+6	+2	4
2	240	243	+3	−1	1
3	210	210	0	−4	16
4	184	189	+5	+1	1
5	190	200	+10	+6	36
6	181	191	+10	+6	36
7	156	150	−6	−10	100
8	220	226	+6	+2	4
9	210	220	+10	+6	36
10	165	163	−2	−6	36
11	188	192	+4	0	0
12	214	216	+2	−2	4

$$\sum D = 48 \qquad\qquad \sum(D - \bar{D})^2 = 274$$
$$\bar{D} = 48/12 = 4$$

2. Next, we determine the mean difference. This yields $\bar{D} = 4$.

3. Since we need to compute S_D, the standard deviation of the differences, we now obtain the deviation of each difference from $\bar{D}$, and then proceed to square each deviation. This yields the column $(D - \bar{D})^2$, and summing this column yields $\sum(D - \bar{D})^2$. The standard deviation of the differences can now be computed by applying a familiar formula with the symbols changed,

$$S_D = \sqrt{\frac{\sum(D - \bar{D})^2}{n - 1}} = \sqrt{\frac{274.00}{11.00}} = 4.99;$$

or we could have used the machine formula,

$$S_D = \sqrt{\frac{\sum D^2 - [(\sum D)^2/n]}{n - 1}}$$

4. Now we compute the standard error of the mean difference, $S_{\bar{D}}$, by

$$S_{\bar{D}} = \frac{S_D}{\sqrt{n}} = \frac{4.99}{3.46} = 1.44.$$

5. Now that we have the standard error of the sampling distribution, we can place our sample statistic, $\bar{D} = 4$, on this distribution by computing t. This is done by

$$t = \frac{\bar{D} - 0}{S_{\bar{D}}} = \frac{4 - 0}{1.44} = 2.78.$$

The Decision

We now enter Table III at $(n - 1)$, or 11 degrees of freedom. You will note that, in the matched-pair situation, we lose only one degree of freedom because only one parameter estimate, $\bar{D}$, was used in the computation of the standard error. To put it another way, if you look back at Table 6.2, you will see that we are actually dealing with a single distribution of 12 numbers, that is, the distribution of $\bar{D}$'s. Assuming a two-tailed test with 11 degrees of freedom, we find that the critical value at the 0.05 level is 2.201. Our obtained t-value of 2.78 obviously exceeds this critical value, and we are therefore able to reject the null hypothesis beyond the 0.05 level. You will note, however, that we do not exceed the 0.01 level.

We are once again in the happy position where we can point to the rejection of the null hypothesis as evidence that supports our contention that the alternate hypothesis is true. In other words, we can now state that we have experimental evidence that epinephrine causes a change in the level of blood cholesterol in rats.

Although this may be getting tiresome, look once again for the word "prove" in the preceding statements. This apparent neurotic preoccupation with the avoidance of that word and what it implies is not without good reason, since it is a word that is used much too loosely by people who should know better. By now you should be aware that one can never "prove" an alternate hypothesis by rejecting a null hypothesis, because there is always the possibility, however small, that the null hypothesis should *not* have been rejected. On the other hand, you may be wondering how this approach can be consistent with the demands of the real world. In other words, does not the ever-present possibility of a Type I error paralyze us into inaction?

It may help to think of it this way. Suppose we have 100 people on the roof of a twenty-story building, and suppose that we wish to test the hypothesis that every time an individual steps off the edge he or she will fall screaming to the pavement below. We take careful notes as we watch the 100 subjects plunge one by one to an untimely death, and now it is our turn. At this point we decide to accept as a working hypothesis the contention that stepping off the tops of tall buildings produces disastrous results! In other words, there must be a practical point where we are convinced by suitable evidence to at least *tentatively* accept the notion that a certain vaccine is effective, or that drug A is more effective than drug B, or that a certain trait is inherited by a particular genetic mechanism. The word "tentative" is the key, because it says that scientific "truths" are not immutable, and that we must be prepared to discard established concepts if additional evidence suggests that they should be discarded.

PROBLEMS

6.1 A pair of dice is rolled 180 times, and a total of 40 sevens appear. On the basis of these data, would you conclude that the dice are honest? (See Section 6.2.)

6.2 A cross between a vestigial-winged fruit fly (vv) and a heterozygous normal-winged fly (Vv) resulted in 20 offspring, 12 of which were of the vestigial phenotype. By using the normal-curve approximation, determine the probability of obtaining 12 or more vestigial-winged flies from such a cross. What is your decision concerning the validity of the genetic model? (See Section 6.2.)

6.3 Serum haptoglobin is known to have an approximate mean value of 100 mg/100 ml in the normal population, with a population standard deviation of 40 mg/100 ml. A sample of 225 cancer patients is found to have a mean haptoglobin value of 113 mg/100 ml. Determine whether the sample is drawn from a group that has a higher haptoglobin concentration than the general population. (See Section 6.3.)

6.4 The mean level of prothrombin in the normal population is known to be approximately 20 mg/100 ml of plasma. A sample of 625 patients showing a vitamin K deficiency has a mean prothrombin level of 18.50 mg/100 ml. The sample standard deviation is 4 mg/100 ml. Do patients with vitamin K deficiency have a significantly lower prothrombin level than that of the general population? (See Section 6.3.)

6.5 A sample of males was drawn from each of two geographically isolated populations of *Rana pipiens*, and their body lengths were measured to the nearest millimeter. From the data below, determine whether there is a statistically significant difference between the males of the two populations in terms of body length (see Section 6.7):

$$\bar{X}_1 = 74 \qquad \bar{X}_2 = 78$$
$$S_1^2 = 225 \qquad S_2^2 = 169$$
$$SS_1 = 2475 \qquad SS_2 = 2535$$
$$n_1 = 12 \qquad n_2 = 16$$

Medium I	Medium II
$\overline{X}_1 = 5.20$	$\overline{X}_2 = 7.50$
$S_1^2 = 16$	$S_2^2 = 25$
$SS_1 = 627$	$SS_2 = 975$
$n_1 = 40$	$n_2 = 40$

6.6 Two different food media were compared in order to determine whether there was any difference in effect on the length of the larval stage in *Drosophila*. Analyze the data above for significant difference between the two media. The length of the larval stage is given in days. (See Section 6.7.)

6.7 A drug which was believed to hasten blood-clotting time was tested by comparing a drug group with a placebo group. Analyze the following data in order to determine whether the mean clotting time of the drug group is significantly lower than the mean clotting time of the placebo group. The clotting time is given in minutes. (See Section 6.7.)

Drug	Placebo
$\overline{X}_D = 4.90$	$\overline{X}_P = 7.45$
$S_D^2 = 10.24$	$S_P^2 = 12.96$
$SS_D = 194.56$	$SS_P = 246.24$
$n_D = 20$	$n_P = 20$

6.8 A sample of 9 measurements yields a mean of 45 and a standard deviation of 8. Estimate the population mean with 95% confidence (See Section 6.6).

6.9 A sample of 14 measurements yields a mean of 50 and a standard deviation of 5. Estimate the population mean with 99% confidence (see Section 6.6).

6.10 An estimation based on a sample of 16 measurements yields a 95% confidence interval of 26.80–33.20. What is the standard deviation of the sample? (See Section 6.6.)

6.11 A group of mice were placed in a series of stress situations which elicited a fear response. After a period of time under these conditions, the mice were compared to those of a control group which had not been put under stress. Analyze the following data to determine whether a significant difference in adrenal-gland weight exists between the two groups. The adrenal weight is expressed in milligrams (see Section 6.7).

Experimental	Control	Experimental	Control
3.8	4.2	3.9	3.6
6.8	4.8	5.9	2.4
8.0	4.8	6.0	3.2
3.6	2.3	5.7	4.9
3.9	6.5	5.6	
4.5	4.9	4.5	

6.12 An investigator tests a drug which he has reason to believe will increase hemoglobin content in grams/100 ml. The hemoglobin content of eight subjects is measured before and after administration of the drug. Analyze the following data in terms of the effectiveness of the drug (see Section 6.8):

Subject	Before	After	Subject	Before	After
1	10	12	5	8	9
2	9	11	6	7	10
3	11	13	7	12	12
4	12	14	8	10	14

6.13 Two groups of plants were used to test the effect of an auxin on height. One group was treated with the auxin and the other group, grown under identical conditions, was left untreated as a control. On the basis of the following data, test the hypothesis that no significant difference in height exists between the experimental and control groups. (See Secton 6.7.)

Experimental	Control
$N_E = 12$	$N_C = 15$
$\overline{X}_E = 260$ mm	$\overline{X}_C = 250$ mm
$SS_E = 800$	$SS_C = 460$

6.14 Two varieties of peas were compared in terms of ascorbic acid content, measured in milligrams/100 g. From the following data, which were derived from ten samples drawn from each variety, determine whether a significant difference in ascorbic acid content exists between the two varieties (see Section 6.7).

Variety I	Variety II	Variety I	Variety II
39	42	28	34
40	39	26	27
34	41	21	25
32	36	19	31
29	28	22	22

6.15 The effect of a nutrient solution on plant growth was tested using 12 plots, each plot containing two plants. In each plot, one plant was treated with the solution and the other plant was left untreated as a control. Analyze the following data to determine whether the treated plants show significantly greater height than the untreated plants. Height is recorded in centimeters. (See Section 6.8.)

Plot	Treated plants	Untreated plants	Plot	Treated plants	Untreated plants
1	24.8	22.6	7	22.4	19.5
2	21.6	21.0	8	26.7	21.6
3	27.8	23.4	9	23.8	20.3
4	29.9	27.5	10	22.8	18.5
5	30.0	39.0	11	26.6	26.0
6	23.0	20.0	12	24.0	21.0

6.16 Two different methods were used to determine the concentration of pro-thrombin in plasma. Both determinations were made on the same subject, using eight subjects in all. On the basis of the following data, where prothrombin is expressed in milligrams/100 ml, determine whether a significant difference exists between the two methods (see Section 6.8):

Subject	Method I	Method II	Subject	Method I	Method II
1	17	18	5	22	23
2	17	17	6	17	15
3	18	20	7	23	25
4	21	24	8	23	22

CHAPTER SEVEN

Enumeration Data— Chi-Square and Poisson Distributions

7.1 ENUMERATION DATA

As the term implies, *enumeration data* result from a counting process. This usually involves assigning experimental units to specific categories in accordance with certain attributes, and then taking counts on the units in each category. This is obviously different from measurement data, which, ideally, can take any value at all between two extremes on a continuum.

To illustrate, suppose that we administer a therapeutic drug to twenty patients suffering from a disease that is usually fatal. Discounting the added possibility of varying degrees of improvement, there are two possible alternatives that can happen in each case: either the patient will recover, or he will not recover. Counting the number of patients who recover and the number who die yields an example of enumeration data.

There are many examples of these "*A*, non-*A*" kinds of data. People either have blue eyes or they do not, a fruit fly either has vestigial wings or it has normal wings, and so on. There are many situations in biological science, especially in genetics, where experimental data are of this enumeration variety.

7.2 GOODNESS-OF-FIT CHI-SQUARE TESTS

One of the most familiar and useful statistical tests is the technique known as chi-square (χ^2). In order to illustrate how this technique can be used to treat enumeration data, we shall return briefly to the coin problem of Section 6.2.

A review of this problem shows that, based on 100 tosses of an unbiased coin, the probability of obtaining 65 or more heads by chance alone was only 0.0019. We therefore rejected the null hypothesis of "no difference" between 65 heads and the expected 50 heads, and concluded that the coin was not fair, but biased in favor of heads. In Section 6.2 this decision was based on the use of the normal-curve approximation to the binomial.

Now we shall use chi-square to determine whether 65 heads and 35 tails is a significant departure from the 50 heads and 50 tails that one would expect from 100 tosses of an honest coin. By now, we are much too sophisticated about these things to really expect *exactly* 50 heads and 50 tails every time a coin is tossed 100 times but, based on the probability value of $\frac{1}{2}$ that is attached to a fair coin, we should certainly be suspicious if the deviation of the observed frequencies is unusually large!

Note that the *observed* frequencies are those actually derived from observations made on experimentally produced data. The *expected* frequencies, on the other hand, are based on some preconceived notion, or *a priori hypothesis*, which in this case is simply the fact that the probability of an honest coin coming up heads in any single toss is $\frac{1}{2}$. Thus the expected frequencies in our case in point are 50 heads and 50 tails.

These observed and expected frequencies are now organized in the form of a table:

	Heads	*Tails*	
Observed	65	35	100
Expected	50	50	100
	115	85	

It may be seen that in this 2 × 2 table we could enter any value in any given cell, but the values to be entered in the three remaining cells would then be fixed by the marginal totals. In other words, one and only one cell frequency may be entered with any freedom as to its value. We therefore say that a 2 × 2 chi-square table has *one* degree of freedom. Thus, like the *t*-test, the chi-square test is based on a family of distributions, where the shape of the distribution is based on degrees of freedom. In general, degrees of freedom associated with chi-square may be determined by the formula

$$(\text{rows} - 1)(\text{columns} - 1) = \text{degrees of freedom},$$

and since a 2 × 2 table has 2 rows and 2 columns, the degrees of freedom are computed by $(2 - 1)(2 - 1) = 1$.

The chi-square formula is now applied to the data as they appear in the table. This formula involves a *summation of the squared absolute differences between each observed frequency and its associated expected frequency, divided by the expected frequency*. Thus

$$\chi^2 = \sum \frac{(O - E)^2}{E}.$$

It should be noted that the size of the obtained chi-square value will be determined by the magnitude of the differences between the observed and expected frequencies. Large differences will produce a large value of chi-square; smaller differences will produce smaller chi-square values, and if no differences exist, then $\chi^2 = 0$.

In our present example, we will need to slightly modify the basic formula. Chi-square is based on a discrete, not a continuous variable. We therefore need to "correct for discontinuity" by subtracting 0.50 from the absolute difference between each observed and expected frequency combination. This is called the Yates correction factor, and is to be used with 2 × 2 tables only. It should also be mentioned here that, although the literature is confusing on this point, it would be well to use the Yates correction factor with *all* 2 × 2 table situations, regardless of the size of cell values.

In our present examples we therefore compute chi-square by

$$\chi^2 = \frac{(|O - E| - 0.50)^2}{E}$$

$$= \frac{(|65 - 50| - 0.50)^2}{50} + \frac{(|35 - 50| - 0.50)^2}{50}$$

$$= \frac{14.5^2}{50} + \frac{14.5^2}{50} = 8.40.$$

If we enter Table V (located in the appendix) at the row associated with one degree of freedom, we move across the chi-square values to 6.635. This is the critical value for the 0.01 level. In the next column we find 7.879, which is the critical value for the 0.005 level. Our obtained value of 8.40 is therefore significant well beyond the 0.005 level, and it appears to correspond to the probability of 0.0019 obtained in Section 6.2 by normal-curve approximation. The probability that this large a difference between the observed and expected frequencies could have occurred by chance alone is therefore so small that we once again suspect that our coin is biased in favor of heads.

In the following example, the *a priori* hypothesis, on which the expected frequencies are based, is derived from a genetic model.

Example. There is a genetic model which assumes that black coat color in mice is inherited as a simple dominant trait, and that brown color is inherited as a recessive trait. A cross between pairs of heterozygous black mice produced an F_2 generation consisting of 220 black mice and 60 brown mice.

1. According to our genetic model, a cross between heterozygous black mice would produce offspring as follows:

$$Bb \times Bb$$

$$F_2: \quad BB, \quad Bb, \quad Bb, \quad bb$$

which represents a phenotype ratio of three black mice to one brown.

2. Now, if the total of 280 offspring occurred in exactly a $3:1$ phenotype ratio, as *expected* from our genetic model, we would have $\frac{3}{4}$ (280) and $\frac{1}{4}$ (280), or 210 black and 70 brown mice. The *observed* ratio of $220:60$ obviously differs from the expected, but is this difference large enough to be significant? In other words, can we accept this difference as being due to chance alone, or is it large enough to lead us to suspect a causative factor, which might possibly cast doubt on the validity of the original genetic model?

3. Our next step involves setting up a chi-square table as follows:

	Black	*Brown*	
Observed	220	60	280
Expected	210	70	280

4. Note that the row totals must be equal. Now, using the Yates correction factor, we apply the chi-square formula:

$$\chi^2 = \frac{(|220 - 210| - 0.50)^2}{210} + \frac{(|60 - 70| - 0.50)^2}{70}$$

$$= \frac{9.5^2}{210} + \frac{9.5^2}{70} = 0.42 + 1.28 = 1.70.$$

5. If we look at Table V with one degree of freedom, we find that a chi-square value of 1.70 is too small to indicate significance at the 0.05 level.

6. We have therefore failed to demonstrate that the observed offspring ratio was significantly different from the expected ratio. We therefore conclude that the observed difference was due to chance alone and we have no evidence to doubt the validity of the genetic model on which the expected frequencies were based.

When using chi-square, we are not limited to only two categories. This is illustrated by the following example.

Example. Suppose that two dihybrids are crossed in a situation where complete dominance is assumed. It is further assumed that no linkage or other complicating factors are present. We therefore have the genetic model

$$AaBb \times AaBb$$

$$F_2: \quad 9 \text{ A–B–}, \quad 3 \text{ A–bb}, \quad 3 \text{ aaB–}, \quad 1 \text{ aabb}$$

which is the classic $9:3:3:1$ phenotype ratio. Now, suppose that the actual F_2 generation shows frequencies of 85 A–B–, 28 A–bb, 35 aaB–, and 12 aabb. Is this result significantly different from the expected frequencies as dictated by the genetic model?

1. As the first step, we set up the chi-square table as follows:

	A–B–	A–bb	aaB–	aabb	
Observed	85	28	35	12	160
Expected	90	30	30	10	160

2. In this case, since we have a chi-square table larger than 2×2, we apply the chi-square formula *without* the Yates correction factor. Thus

$$\chi^2 = \frac{(85-90)^2}{90} + \frac{(28-30)^2}{30} + \frac{(35-30)^2}{30} + \frac{(12-10)^2}{10}$$

$$= \frac{25}{90} + \frac{4}{30} + \frac{25}{30} + \frac{4}{10} = 1.63.$$

3. Since our chi-square table has four columns and two rows, we enter Table V at $(4-1)(2-1)$, or three degrees of freedom. With three degrees of freedom, the critical value at the 0.05 level is 7.815. Since our obtained chi-square value does not equal or exceed this critical value, we have failed to provide statistical evidence that the assumed genetic model is not operating as expected.

The following example illustrates the use of chi-square as a tool in genetic detective work when establishing genetic models.

Example. In fowls, the creeper gene (producing deformed legs) is dominant over the gene for normal leg development. A series of crosses between heterozygous creepers (Cc) produce a phenotype ratio of 164 creepers to 76 normal birds.

1. Simple inspection of the obtained phenotype ratio reveals an obvious deviation from the $3:1$ ratio expected for a Cc $\times$ Cc cross. Is it different enough to lead us to suspect a causative factor which would render the $3:1$ model invalid? Application of the chi-square test shows:

	Creeper	Normal	
Observed	164	76	240
Expected	180	60	240

$$\chi^2 = \frac{(|164 - 180| - 0.50)^2}{180} + \frac{(|76 - 60| - 0.50)^2}{60}$$

$$= \frac{15.5^2}{180} + \frac{15.5^2}{60} = 1.33 + 4.00 = 5.33.$$

2. Entering Table V at the level of one degree of freedom, we find our chi-square value of 5.33 to be significant beyond the 0.05 level. We now have statistical support for the suspicion that the observed genetic ratio does not conform to the $3:1$ model.

3. Inspection of the $164:76$ observed ratio of creepers to normal birds suggests a proximity to a $2:1$ model. We therefore test this new idea with chi-square by making a new series of crosses between heterozygous creepers (Cc), from which we obtain 134 creepers and 58 normal birds.

	Creeper	Normal	
Observed	134	58	192
Expected	128	64	192

$$\chi^2 = \frac{(|134 - 128| - 0.50)^2}{128} + \frac{(|58 - 64| - 0.50)^2}{64}$$

$$= \frac{5.5^2}{128} + \frac{5.5^2}{64} = 0.709.$$

4. Table V shows that a chi-square value of 0.709 is not significant, and we may conclude that the observed ratio fits a $2:1$ model much better than it conforms to a $3:1$ model.

5. On this basis, we now look for a genetic mechanism other than that of simple dominance with no complicating factors. A look at the observed ratio in step (1) suggests that if 76 homozygous recessive (normal) birds

were produced, 3×76, or 228 creepers should have been produced. In other words, from an uncomplicated cross between Cc and Cc we should, theoretically, obtain CC, Cc, Cc, and cc as the F_2 generation. Now, if we were to assume that CC is lethal in the homozygous condition, this might account for obtaining what appears to be Cc and cc genotypes only, and could therefore explain the missing creepers. As a matter of fact, further investigation would confirm this lethal-gene hypothesis.

The foregoing example once again reminds us of the paramount importance attached to the investigator's knowledge of the *biological* aspects of a problem. The statistician can provide us with tools of inference, but the important answers will always be provided by insights derived from familiarity with the principles. It has been said that research does not solve problems; research provides data with which the trained mind can then attack a specific problem.

7.3 CHI-SQUARE WITHOUT *A PRIORI* HYPOTHESIS

In the preceding section we dealt with situations where some kind of expected frequencies were either known or assumed. Now we will need to consider the application of chi-square techniques to cases in which there are no *a priori* expected frequencies. The following example will illustrate the basic procedure in a simplified fashion by the use of contrived values.

Example. Suppose that we wish to know whether an association exists between the factors sex and hair color. We proceed to check the first 50 men and the first 50 women who come down the street, noting in each case whether the individual is blonde or brunette. To keep our illustration simple, we will ignore the question of whether the people are blonde as a result of hereditary or environmental influences.

1.   As the first step, we will organize the data in the form of Table 7.1:

TABLE 7.1

Sex	Hair color		
	Blonde	Brunette	
Men	20	30	50
Women	24	26	50
	44	56	100

This is called a contingency table, and we are, in effect, asking whether hair color is *contingent* upon sex.

2.   Note that we have no *a priori* expected frequencies. We must therefore compute the expected frequencies, using the marginal totals as a basis for the calculations. The rationale underlying the computation of expected frequencies is a form of null hypothesis wherein we assume that hair color is *not* contingent upon sex. If this is true, then we should observe no tendency for one sex or the other to contain a preponderance of blondes (or brunettes). Since we have established a table with nice round figures, it may easily be seen that $\frac{1}{2}$ the total sample are men and $\frac{1}{2}$ are women. Therefore, if *no* special relationship exists between sex and hair color, we could *expect* that $\frac{1}{2}$ the blondes in the sample would be men and $\frac{1}{2}$ would be women! The same reasoning then applies to the brunettes. Working with the marginal totals, we therefore have

$$\frac{50}{100} \times 44 = 22, \qquad \frac{50}{100} \times 44 = 22,$$
$$\frac{50}{100} \times 56 = 28, \qquad \frac{50}{100} \times 56 = 28.$$

The completed table, with computed expected frequencies, then becomes

	Blonde	Brunette	
Men	20 (22)	30 (28)	50
Women	24 (22)	26 (28)	50
	44	56	100

3.   Now the chi-square value is computed, and since we are dealing with a 2×2 table, we will use the Yates correction factor. Thus

$$\chi^2 = \frac{(|20 - 22| - 0.50)^2}{22} + \frac{(|30 - 28| - 0.50)^2}{28}$$
$$+ \frac{(|24 - 22| - 0.50)^2}{22} + \frac{(|26 - 28| - 0.50)^2}{28}$$
$$= 0.102 + 0.080 + 0.102 + 0.080 = 0.364.$$

4.   Entering Table V at one degree of freedom, we find that 0.364 is not significant at the 0.05 level. We have therefore not shown that the observed frequencies and those computed on the basis of a "no relationship" hypothesis are significantly different, and we are led to the conclusion that hair color is not associated with sex. In other words, the two factors appear to be independent.

As before, this form of chi-square test may be used where more than two categories are involved, as in the following example.

Example. A therapeutic drug was tested against a placebo in terms of three subjectively evaluated patient categories: (1) much improved, (2) slightly improved, and (3) not improved. A total of 120 patients were assigned to the drug group and 90 other patients were given the placebo. All were judged to be in approximately the same initial condition. Physician evaluation was then made without knowing which treatment the patient received. The resulting data were organized in the following 2 × 3 table:

	Much improved	*Slightly improved*	*Not improved*	
Drug	60	32	28	120
Placebo	28	17	45	90
	88	49	73	210

1. This time, we base our calculations of expected frequencies on the hypothesis that no significant difference in degree of improvement exists between the drug and the placebo groups. Therefore since $\frac{120}{210}$ of the sample consists of the drug group, we should expect $\frac{120}{210} \times 88$ to yield the expected frequency associated with "drug—much improved," and so on. The computed expected frequencies are therefore:

Drug—Much improved: $\frac{120}{210} \times 88 = 50.28$
Placebo—Improved: $\frac{90}{210} \times 88 = 37.71$

Drug—Slightly improved: $\frac{120}{210} \times 49 = 28.00$
Placebo—Slightly improved: $\frac{90}{210} \times 49 = 21.00$

Drug—Not improved: $\frac{120}{210} \times 73 = 41.71$
Placebo—Not improved: $\frac{90}{210} \times 73 = 31.28$

We may now organize the data in the form of columns as follows:

$O - E$	$(O - E)^2$	$(O - E)^2/E$
60–50.28	94.48	1.88
32–28.00	16.00	0.57
28–41.71	187.96	4.51
28–37.71	94.28	2.50
17–21.00	16.00	0.76
45–31.28	188.24	6.02
		$\chi^2 = 16.24$

Entering Table V with $(3 - 1)(2 - 1)$, or two degrees of freedom, we find that a chi-square value of 16.24 is significant beyond the 0.005 level. We therefore have statistical evidence that a significant difference in degree of improvement does indeed exist between the placebo group and the drug group.

7.4 CAUTIONS IN THE USE OF CHI-SQUARE

It is important to remember that only *frequency* data may be analyzed by chi-square. It is not appropriate to apply this technique to data existing in the form of percentages or proportions. In many cases, however, it is possible to convert proportions to frequencies and chi-square may then be applied.

We have already discussed the wisdom of using the Yates correction factor in all 2×2 table analyses. Even with large samples, the difference between chi-square values obtained with and without this correction factor may be important enough to affect values of borderline significance.

Finally, it is important to note that each cell in a chi-square table should contain a minimum *expected frequency* of 5. Tables containing expected frequencies of less than five may produce results of doubtful validity, though there is some disagreement among authors as to whether 5 is the "magic number." Discussions of how to handle situations where one or more expected frequencies are less than 5 are found in references listed in the bibliography, but the best approach is to try to avoid such problems by careful planning when designing experiments.

7.5 THE POISSON DISTRIBUTION

Suppose that we survey a certain stretch of highway in order to determine the incidence of dead wildlife found by the roadside and presumably killed by automobiles. Suppose further that, as we go along the highway, we glance at any randomly selected spot beside the road. What is the probability of finding a dead animal at that specific location? Assuming that the roadside is not littered with dead animals, it should be obvious that the probability is exceedingly small—so small, in fact, as to approach zero! In other words, the probability (p) of finding an animal in a specific location is very small compared to the probability (q) of *not* finding an animal in the same specific location.

In cases such as this it makes no sense to count "A's" and "non-A's." For example, how many "nonfish" would be located in a pond if we assume that a "nonfish" is simply a space where a fish is not found? Or, how many "nonclicks" are heard from a Geiger counter in a unit of time?

In situations such as these we are dealing with isolated events as they occur in time or space. The number of clicks produced in a unit of time, the number of tadpoles in a pond, and the number of yeast cells per cubic millimeter in a suspension are examples of situations involving a certain kind of isolated event in a time or space continuum.

Since p is therefore very small and q is very large, application of the binomial distribution is not appropriate, and we therefore need to consider the *Poisson* distribution.

The Poisson distribution is based on a relationship obtained from the calculus. It may be shown that if we have a series,

$$\frac{1}{0!} + \frac{1}{1!} + \frac{1}{2!} + \frac{1}{3!} + \frac{1}{4!} + \cdots,$$

and if the members of this series are summed, we obtain a number approximating 2.718. This number is given the symbol e and is an important value in mathematics, since it is the base of the natural, or Naperian logarithms. It is similar to pi in that it is an apparently endless, nonrepeating decimal.

Now, if e is raised to any power, which we will symbolize by λ (lambda), $e^{\lambda} = 2.718^{\lambda}$ and therefore

$$e^{\lambda} = \frac{\lambda^0}{0!} + \frac{\lambda^1}{1!} + \frac{\lambda^2}{2!} + \frac{\lambda^3}{3!} + \frac{\lambda^4}{4!} \cdots$$

Since $e^{\lambda} \times e^{-\lambda}$ is equal to 1, and since e^{λ} is equal to the above series, we may substitute the above series for e^{λ} and obtain

$$1 = e^{-\lambda} \times e^{\lambda} = e^{-\lambda}\left(1 + \lambda + \frac{\lambda^2}{2!} + \frac{\lambda^3}{3!} + \frac{\lambda^4}{4!} \cdots\right)$$

$$= e^{-\lambda} + e^{-\lambda}\lambda + \frac{e^{-\lambda}\lambda^2}{2!} + \frac{e^{-\lambda}\lambda^3}{3!} + \frac{e^{-\lambda}\lambda^4}{4!} \cdots$$

which is the Poisson distribution, and which describes the occurrence of an isolated event in a time or space continuum; λ is used to represent the *observed mean occurrence* of that event.

It should be noted that $e^{-\lambda}$ represents the probability of zero events, $e^{-\lambda}\lambda$ the probability of one event, $e^{-\lambda}\lambda^2/2!$ the probability of two events, and so forth.

7.6 VARIANCE OF THE POISSON DISTRIBUTION

It will be recalled that the mean and variance of the binomial distribution $(p + q)^n$ are respectively,

$$\mu = np \quad \text{and} \quad \sigma^2 = npq.$$

Now, it may be seen that if p approaches zero, and q therefore approaches 1, the variance of the Poisson distribution is practically $np(1)$, or np, and therefore approximates the mean. The mean and variance values of the Poisson distribution are thus considered to be equal, and this is a useful and simplifying aspect of this distribution.

7.7 DETERMINING THE PROBABILITY OF ISOLATED EVENTS

As indicated previously, the Poisson distribution may be used to determine the probabilities of 0, 1, 2, 3, etc., events occurring in a given continuum. Consider the following example.

Example. A survey was made along a 20-mile stretch of highway to determine the prevalence of dead wildlife found on or near the road. The total number of dead animals counted was 15. Assuming the distribution over the 20-mile distance to be relatively uniform, $\overline{X}$ is computed as our sample estimate of λ.

$$\overline{X} = \frac{15}{20} = 0.75 \text{ animals/mile.}$$

What is the probability of finding no dead animals in any randomly selected mile along this stretch of highway? What is the probability of finding one animal? Two animals?

1. Since $\overline{X} = 0.75$, we will first need to determine the value of $e^{-\overline{X}}$. This may be done by consulting Table VI in the appendix, where we find that for $\lambda = 0.75$, $e^{-\overline{X}} = 0.472$.

2. Since the probability for zero events is given by the Poisson distribution as $e^{-\overline{X}}$, the probability of finding *no* animals in a one-mile stretch is 0.472.

3. The probability of finding one animal is found by $e^{-\overline{X}}\overline{X}$. Therefore,

$$(0.472)(0.75) = 0.354.$$

4. Since the probability of finding two animals is given by $e^{-\overline{X}}\overline{X}^2/2!$, we have

$$\frac{(0.472)(0.563)}{2 \times 1} = 0.133.$$

5. The probabilities associated with finding three, four, five, or more animals may be easily found by using the corresponding values in the Poisson distribution.

7.8 THE CONFIDENCE INTERVAL OF A COUNT

In Chapter 5 we saw that sample statistics can be used to estimate a population parameter with certain confidence limits. It will be recalled that the general method for such estimation involves the relationship

$$\overline{X} \pm 1.96(S_{\overline{X}}).$$

Using the Poisson distribution, we make estimations involving counts, again using a sample statistic as a basic for inference about a population value. Such estimations might be applied to a red-blood-cell count in terms of cells per cubic millimeter, or to bacterial counts, yeast cell counts, etc. An estimation of the density of a certain type of organism in a pond is another example.

Naturally, estimations of organism density assume a uniform distribution. This underlines the importance of careful sampling as well as careful counting techniques.

The following example will illustrate the procedure of estimation, using the Poisson distribution.

Example. A sample of blood was removed from a patient and diluted $1:200$. A total of 250 platelets were counted in a counting chamber having a volume of 0.2 mm^3. What is the 95 percent confidence interval for the patient's true platelet count per cubic millimeter?

1. The sample count yielded a total of 250 platelets/0.2 mm; this was based on a blood sample which had been diluted $1:200$. Our sample count is therefore 250, and we will symbolize this by C.

2. Keeping in mind that the variance of the Poisson distribution is equal to the mean, we can find the standard deviation of C by $\sqrt{C}$.

3. We then have:

$$C \pm 1.96\sqrt{C} = 250 \pm 1.96\sqrt{250},$$

$$250 \pm 1.96(15.81) = 250 \pm 30.99,$$

$$219.01\text{--}280.99/0.2 \text{ mm}^3.$$

4. Since the blood sample represented a $1:200$ dilution, we need to multiply the above figures by 200 and then multiply the product by 5 to obtain the estimated true platelet count per cubic millimeter. Thus we have

$$(219.01\text{--}280.99) \times 200 \times 5,$$

$$219,010 \le \lambda \le 280,990$$

as the 95-percent confidence-limit estimate of the true platelet count per cubic millimeter in the patient's circulatory system.

7.9 DETERMINING RANDOMNESS

A useful application of the Poisson distribution involves the determination of randomness. An ecologist may wish to determine whether a particular plant is distributed randomly over a given area, or whether the plant has a tendency toward clumping. A bacteriologist may wish to know whether bacterial growth on plates is random, or whether unusual growth has occurred on certain plates. A radiation biologist may want to determine whether the counts produced by a Geiger counter are random, since nonrandomness might indicate a defective instrument.

These and other problems may be approached with a method that combines the Poisson distribution with an application of chi-square. The following example illustrates the procedure.

Example. An ecologist wants to determine whether goldenrod plants containing one or more galls tend to be randomly distributed or clumped. He reasons that clumping might suggest that the wasps that produce the galls range only short distances from the point of emergence. He therefore selects a field in which goldenrod plants are uniformly distributed and randomly throws 100 quadrats throughout the field. Counting the number of gall-bearing plants in each quadrat results in the following data:

> 44 quadrats contain zero plants with galls,
> 24 quadrats contain one plant with galls,
> 18 quadrats contain two plants with galls,
> 14 quadrats contain three plants with galls.
> _____
> 100 quadrats, total.

1. Multiplying the number of quadrats times the number of plants per quadrat, we have:

$$44 \times 0 = 0,$$
$$24 \times 1 = 24,$$
$$18 \times 2 = 36,$$
$$14 \times 3 = 42,$$

yielding a total of 102 plants counted over 100 quadrats.

2. Thus, our *sample estimate* of λ is $\overline{X} = 102/100$, or 1.02 plants per quadrat.

3. Table VI indicates that when $\overline{X} = 1.02$, $e^{-\overline{X}}$ is 0.361.

4. Our next step is to determine the "Poisson probabilities." In other words, we need to find the probability that any given quadrat thrown at random will contain zero plants with galls, the probability it will contain one plant with galls, and so on. Keep in mind that these are the probabilities that would occur *if* the plants with galls are randomly

distributed. Thus

$$P_{\text{Zero plant}} = e^{-\overline{X}} = 0.361;$$

$$P_{\text{One plant}} = e^{-\overline{X}}\overline{X} = (0.361)(1.02) = 0.368;$$

$$P_{\text{Two plants}} = \frac{e^{-\overline{X}}\overline{X}^2}{2 \times 1} = \frac{(0.361)(1.04)}{2} = 0.188;$$

$$P_{\text{Three plants}} = \frac{e^{-\overline{X}}\overline{X}^3}{3 \times 2 \times 1} = \frac{(0.361)(1.06)}{6} = 0.063.$$

5. Now we can determine the numbers of quadrats that would be expected to contain zero, one, two, or three plants with galls if the plants with galls are randomly distributed. To do so, we simply multiply each probability value obtained in step (4) by the total number of quadrats (100). Thus,

$$100 \times 0.361 = 36.1,$$

$$100 \times 0.368 = 36.8,$$

$$100 \times 0.188 = 18.8,$$

$$100 \times 0.063 = 6.3,$$

6. Next, we use a chi-square "goodness-of-fit" test to determine whether the differences between the observed and the expected quadrat frequencies occurred by chance alone. Since an organism that is distributed according to the Poisson distribution *is* randomly distributed, we wish to see whether a statistically significant difference exists between the observed frequencies and the frequencies that are *expected* on the basis of the Poisson probabilities. Thus,

	Plants			
	0	*1*	*2*	*3*
Observed	44	24	18	14
Expected	36.1	36.8	18.8	6.3

$$\chi^2 = \frac{(44 - 36.1)^2}{36.1} + \frac{(24 - 36.8)^2}{36.8} + \frac{(18 - 18.8)^2}{18.8} + \frac{(14 - 6.3)^2}{6.3}$$

$$= 15.63.$$

7. We have a 4×2 chi-square table and ordinarily we would therefore enter Table V at three degrees of freedom. In this case, however, our expected frequencies are based on $\overline{X} = 1.02$, which is an *estimate* of λ, the true mean of the population. We therefore must lose another degree of freedom because of the parameter estimate used in our calculations. Entering Table V at two degrees of freedom, we find that a chi-square value of 15.63 is well beyond the critical value for the 0.005 level. We therefore reject the null hypothesis of "no difference" between the observed and expected frequencies. Since the expected frequencies are based on a *random* Poisson distribution, we therefore conclude that the observed frequencies do *not* constitute a random distribution. According to our data, therefore, the plants containing the galls are not randomly distributed.

PROBLEMS

7.1 When pink four-o'clocks are crossed, it is expected that the F_2 frequencies will be on the order of 1 red : 2 pink : 1 white. An investigator obtains an actual ratio of 30 red : 48 pink : 27 white. Are these results consistent with the genetic model? (See Section 7.2.)

7.2 The F_2 generation resulting from crosses between heterozygous red owls contained 16 red and 8 grey owls. Are these results consistent with the genetic theory that red is dominant over grey? (See Section 7.2.)

7.3 A vaccine was developed which was supposed to protect mice against a particularly virulent bacterium. A group of 55 mice was given the vaccine and then challenged with a heavy dose of the bacterium. Another group of 55 mice, which had not been vaccinated, was challenged with the same dose. Twenty-nine mice in the vaccine group contracted the disease, and thirty-five in the control group became ill. Test for effectiveness of the vaccine (see Section 7.3).

7.4 An experiment was performed to determine the degree to which the isopod *P. laevis* can acclimate to different temperatures. Of 50 specimens held at 10°C, 32 died. Meanwhile, of 50 specimens held at 20°C, 20 died. Analyze the data for the difference in their abilities to acclimate to the two temperatures (see Section 7.3).

7.5 A group of 500 Germans living in Germany was analyzed with respect to blood types. Another group of 476 persons of German origin, but living in Hungary, was also analyzed for blood types. Analyze the following data for possible difference between the two groups in terms of blood-type frequencies (see Section 7.3):

Group	AB	A	O	B	*Totals*
I (Germany)	25	215	200	60	500
II (Hungary)	15	207	194	60	476

7.6 An investigator set up an experiment involving the effect of a light gradient on the gastropod *Littorina*. The gradient was established with three zones of light intensity, and of a sample of 33 gastropods, 18 were noted in zone 1, 10 in zone 2, and 5 in zone 3. Test the hypothesis that the light gradient had no effect on the distribution of the organism (see Section 7.2).

7.7 Analyze the following contingency table to determine whether color in a certain variety of mice depends on, or is associated with sex (see Section 7.3).

Color	Male	Female
Black	26	15
Brown	17	24
White	15	18

7.8 A drug believed to have teratogenic properties was given to a group of 65 pregnant female rats, and it was later noted that 23 females of this group produced litters containing at least one malformed offspring. From a group of 85 females used as a control, 12 litters contained at least one malformed offspring. On the basis of these data, what conclusion might be drawn relative to the teratogenic properties of the drug? (See Section 7.3.)

7.9 An antimalarial drug was given to 1500 men, and 15 individuals showed an anaphylactic reaction. Of 1400 women given the same drug, 40 individuals had a similar reaction. Analyze these data to determine whether an association exists between sex and an allergic reaction to the drug (see Section 7.3).

7.10 In testing a new antibiotic, it was found that 5 out of 5000 patients suffered an allergic reaction. On the basis of these data, compute the probability that out of 1000 patients chosen at random,

   a) exactly one would suffer a reaction;
   b) exactly three would suffer a reaction;
   c) more than three would suffer a reaction.

(See Section 7.7.) [*Hint:* $\lambda = np$.]

7.11 In order to determine the mode of distribution of a certain species of cricket, a total of 200 quadrats were randomly thrown on a field. Each quadrat was examined for crickets, and the following data were obtained in terms of the cricket count:

Quadrats	Number of crickets/quadrat
98	0
52	1
28	2
22	3

Analyze these data to determine whether the crickets are randomly dispersed throughout the area (see Section 7.9).

7.12 A Geiger counter yields a total of 10,000 counts during a one-minute period when exposed to a radioactive source. Compute the 95% confidence interval in terms of counts per minute (see Section 7.8).

7.13 An investigator counted 256 yeast cells in a one-milliliter sample taken from a culture. Compute the 95% confidence interval for the mean count per milliliter in the culture (see Section 7.8).

7.14 Fifty-five samples were randomly collected from a pond, using a plankton net, in order to determine the distribution of mayfly larvae. On the basis of the following data, are the larvae randomly distributed over the pond area or are they clumped? (See Section 7.9.)

Number of samples	Larvae per sample
30	0
20	1
13	2
10	3

CHAPTER EIGHT

Analysis of Variance—
Part I

8.1 COMPARISONS AMONG MORE THAN TWO GROUPS

In Chapter 6 we discussed how Student's t-test can be used to analyze data derived from two-group designs. This is a legitimate way to test null hypotheses concerning two groups, but there are many situations in biological research that call for comparisons among three or more groups.

Comparing three or more groups with a t-test presents certain problems. For one thing, if we compared groups A, B, C, and D for significant differences, we would have to test for differences between pairs AB, AC, AD, BC, BD, and CD; this would require *six separate t-tests*. Not only would this be tedious, but what is more important, the possibility of error would increase with the number of tests, bringing the total probability of error to a prohibitive level.

Fortunately, this problem can be overcome with a statistical test called *analysis of variance*, which, for this and other reasons, is one of the most useful statistical procedures presently available for biological research. We will see in the next chapter that analysis of variance allows the use of a variety of sophisticated designs from which we can obtain a maximum of information.

8.2 ANALYSIS OF VARIANCE

The purpose of this section is to develop the principles upon which analysis of variance is based. This will be done as simply as possible, using small amounts of data to avoid getting all tied up in arithmetic.

Suppose that we want to determine whether diets A, B, and C show statistically significant differences in terms of their effects on weight gain in mice. Or, to put it differently, we are interested in whether or not there will be real differences among the theoretical *populations* that could be fed with diet A, diet B, or diet C. By now you should know that in this experiment we are manipulating an independent variable (diet) and observing the effects on a dependent variable (weight gain).

Let's say that we select 15 mice of a certain variety from the general population of that variety and then randomly assign them to the three treatments (diets) so that we have five replications per treatment. The probability is therefore $\frac{1}{3}$ that each mouse in the sample will be assigned to diet A, diet B, or diet C. After a specified period we measure the gain in weight of each mouse and obtain the data shown in Table 8.1.

TABLE 8.1

A	B	C	
32	36	35	
37	38	30	$k = 3$
34	37	36	$n_T = 15$
33	30	29	$\overline{\overline{X}} = 33.5$
30	34	31	$S^2 = 8.8$
$n_A = 5$	$n_B = 5$	$n_C = 5$	
$\overline{X}_A = 33.2$	$\overline{X}_B = 35.0$	$\overline{X}_C = 32.2$	
$S_A^2 = 6.7$	$S_B^2 = 10.0$	$S_C^2 = 9.7$	

Looking at Table 8.1 you will note that n_A, n_B, and n_C represent the number of replications per treatment; in this case they are each five. The total sample (15) is denoted by n_T. Also note that the mean of the three treatment means is represented by $\overline{\overline{X}}$; it is also called the *grand mean*.

The variance *within* each treatment group has been computed by dividing the sum of the squared deviations from each treatment mean by $(n_A - 1)$, $(n_B - 1)$, and $(n_C - 1)$, respectively. Since the treatment n-numbers are all the same, the mean of the group variances is calculated simply as the mean of 6.7, 10.0, and 9.7, which is 8.8. If the n-numbers were not all the same, we would calculate a pooled variance much as we did in Section 6.7 when we were comparing two groups with a t-test. This figure (8.8) is represented by S^2 and it is called the *within groups* or *error* variance. As usual, it is due to the random variation that occurs among the subjects. Finally, there are three treatment groups; we will use k as the general symbol for number of treatments, so in this case $k = 3$.

Our sample of $n_T = 15$ mice was drawn from a population that has a mean (μ) and a variance (σ^2). As usual, we do not know the true values of μ and σ^2, and we can never know them unless we set up an experiment in which we use *all* the mice of that particular variety that currently exist and *could* exist—which is obviously impossible!

On the other hand, there are two items in our sample data that can be used to *estimate* σ^2. First, the within-groups variance is itself an estimate of σ^2. Since our within-groups variance (S^2) is 8.8, this therefore becomes an estimate of σ^2, which is the random variation in the population.

Secondly, the variability, or differences, among the treatment means can be used in a more indirect way to estimate σ^2. We begin with the assumption that we can theoretically draw a billion billion groups of $n = 5$ from the population and construct a sampling distribution of group means. At this point you should recall that the variance of this theoretical distribution of means is $\sigma_{\bar{X}}^2$, and is computed by

$$\sigma_{\bar{X}}^2 = \frac{\sigma^2}{n},$$

where σ^2 is the population variance and n is the number of replications per group, which in our example is five.

We can now easily obtain an estimate of $\sigma_{\bar{X}}^2$ from the three treatment means in Table 8.1. All we have to do is think of the three means as a distribution of three numbers with a mean ($\bar{\bar{X}}$) of 33.5. Thus,

$$\bar{X}_A = 33.2$$
$$\bar{X}_B = 35.0$$
$$\bar{X}_C = 32.2$$
$$\overline{\overline{X}} = 33.5$$

We can now find the sum of squared deviations from the grand mean, or "sum of squares," by

$$(33.2 - 33.5)^2 + (35.0 - 33.5)^2 + (32.2 - 33.5)^2,$$

obtaining 4.02. We now divide the sum of squares (SS) by $(k - 1)$, which by now you recognize as the method of computing an unbiased variance estimate. We therefore have

$$S_{\bar{X}}^2 = \frac{SS}{k - 1} = \frac{4.02}{2} = 2.01.$$

Now, since $S_{\bar{X}}^2 = S^2/n$; it follows that $S^2 = nS_{\bar{X}}^2$, in which n is the number of replications per treatment. Since n_A, n_B, and n_C are all five, in this case $n = 5$. Therefore,

$$S^2 = 5(2.01) = 10.05,$$

which is an estimate of σ^2 obtained from the variability (differences) among the treatment means. It is not the same value as the one obtained from the within-groups variance (8.8), but we have to remember that the estimate provided by the treatment means is based on three numbers while the within-groups variance estimate is based on fifteen numbers.

At this point we are ready to state one of the most important principles of analysis of variance: *Any variability (differences) among the treatment means consists of an estimate of σ^2 plus any extra variability that is due to and originating from differences produced by treatment effects.*

In our case in point, the estimate of σ^2 obtained from the group means is 10.05 and the estimate provided by the within-groups variance is 8.8. We now must question whether the variance estimate derived from the treatment means is only an estimate of σ^2 or whether it contains an extra component due to differences in the means caused by differential effects of the diets.

We do this by using a variance ratio that follows the F-distribution (Table VII). Thus,

$$F = \frac{\sigma^2 + K}{\sigma^2},$$

in which K is the variability in the treatment means *beyond* that which we would expect as an estimate of σ^2 alone. For example, if K is zero, then the F-value would be 1. This would indicate that the variability among the treatment means is only an estimate of σ^2 and does *not* contain an extra component that is due to and originating from treatment effects. On the other hand, we know that some chance-alone variability is likely; the degree of significance therefore depends on the extent to which the *among-group* variance estimate exceeds the *within-group* variance estimate. In our example, we have

$$F_{[2,\,12]} = \frac{\sigma^2 + K}{\sigma^2} = \frac{10.05}{8.8} = 1.14.$$

The subscripts, 2 and 12, represent the treatment and within-groups degrees of freedom, respectively. Referring to Table 8.1, you recall that we found the variance of the treatment means by dividing the sum of squares by $(k - 1)$, or 2. Also, when we computed the within-groups variance, we calculated each treatment group separately by dividing the sum of squares by $(n_A - 1)$, $(n_B - 1)$, and $(n_C - 1)$. The degrees of freedom for within-groups is therefore $4 + 4 + 4$, or 12.

Now turn to the F-table (Table VII) and note that it is set up like a mileage chart on a road map. Since two degrees of freedom are associated with the numerator and 12 with the denominator, we go down the second column to the twelfth row. Where column and row intersect we locate 3.88 and 6.93 as the critical values for the 0.05 and 0.01 levels, respectively. Since we obtained an F-value of 1.14, it clearly does not equal or exceed the critical value (3.88) for the 0.05 level. We must therefore conclude that the variability (or differences) among the diet means is not significantly greater than one would expect it to be if it is only an estimate of σ^2. In other words, we have no evidence that there are significant differences among diets A, B, and C relative to gain in weight.

Now let's take a look at Table 8.2, where we can see that we have the same experiment except that we now have a different set of three diets—D, E, and F.

TABLE 8.2

D	E	F	
29	38	41	
30	33	34	$k = 3$
34	34	40	$n_T = 15$
31	32	42	$\overline{\overline{X}} = 34.8$
27	39	38	$S^2 = 8.8$
$n_D = 55$	$n_E = 5$	$n_F = 5$	
$\overline{X}_D = 30.2$	$\overline{X}_E = 35.2$	$\overline{X}_F = 39.0$	
$S_D^2 = 6.7$	$S_E^2 = 9.7$	$S_F^2 = 10.0$	

We have contrived the data so that we have the same within-groups (error) variance as before (8.8), but when we compute the variance of our treatment means we now have

$$\frac{(30.2 - 34.8)^2 + (35.2 - 34.8)^2 + (39.0 - 34.8)^2}{2},$$

which yields $S_{\overline{X}}^2 = 19.5$. Following the same reasoning as we did with the previous example, we now have

$$S^2 = nS_{\overline{X}}^2 = 5(19.5) = 97.5.$$

Once again, we compute F by

$$F_{[2, 12]} = \frac{\sigma^2 + K}{\sigma^2} = \frac{97.5}{8.8} = 11.07.$$

Consulting Table VII with two and 12 degrees of freedom as before, we find that our F-value of 11.07 is significant well beyond the 0.01 level. We therefore conclude that statistically significant differences exist among the treatment means. Or, to put it another way, the variability of the treatment means provided an estimate of σ^2 *plus an added component* (K), which we assume is due to an additional variability among the treatment means, which is in turn caused by differential effects of the diets.

To summarize, since the total variability consists of components contributed by natural variation (error) and by treatment effects, if any, we need to somehow separate, assess, and compare these two components. If in doing so, we find that the variability *among* the treatment groups is not meaningfully greater than the variability *within* the groups (error), we cannot conclude that the treatment effects are statistically significant. If, on the other hand, the among-group variance is significantly greater than the random error within groups, we may suspect that a major portion of the variability among the groups was produced by differential effects of the treatments.

In this section we have looked at the major principle upon which analysis of variance is based. In a later section of this chapter, we will take a completely randomized design and analyze it with methods that are more mechanical. Meanwhile, if you have found this section difficult or confusing, you would do well to study it again because it is essential that your understanding of this important test go beyond machine formulas and computer programs.

8.3 ASSUMPTIONS RELATED TO ANALYSIS OF VARIANCE

Analysis of variance can be described by a so-called linear model that looks like the following:

$$X = \mu + t + \epsilon$$

or

$$X - \mu = t + \epsilon.$$

A little study of the above model shows that any given measurement (X) differs from the population mean (μ) by an amount that is due to the combined effects of the treatment (t) to which the unit is subjected, and natural variation, or error (ϵ). If the treatment has no effect, then t will be zero and the only source of difference from the true mean is random variation. This corresponds to what we saw in the last section.

There are certain assumptions that must be met if the above model is to hold true and the test results can be considered reliable:

1. It is assumed that the error term is randomly distributed throughout the treatment groups. To increase the probability that this assumption is met, we must carefully randomize the assignment of subjects to treatments. For example, if we were to assign active mice to one treatment and sluggish mice to the other, then the error may not be randomly distributed, but *systematic.*

2. It is assumed that the treatment group variances are *homogeneous.* In other words, they should not differ from each other by an amount greater than one would expect on the basis of chance. This requirement of homogeneous variances is often referred to as *homoscedasticity*, which is a word that never fails to liven up a dull conversation at a cocktail party. Looking back at Table 8.2, and considering the basic principle of analysis of variance, it may be seen that *each* of the treatment variances *should* be and is *assumed* to be a relatively good estimate of σ^2. Quite obviously, they cannot be reliable estimates of σ^2 if they are wildly different. If the variances of the treatment means are not reasonably homogeneous, we might suspect that we were not very successful in our attempt to randomly assign the subjects to the various treatment groups. Or, it may be that certain experimental procedures applied to group B may have produced a variance in group B that is considerably greater or less than that of A, C, etc. In one situation, a bottom-dwelling organism was collected from a lake bed and there were vast differences in the variances within collecting sites. It was discovered that the organism was not randomly distributed but existed in clumps; it was therefore a matter of chance whether the collecting device produced hundreds of the organisms or none at all, with each replication! Another possible source of heterogeneous variances is a multiplicative rather than an additive effect of treatments. Referring to the linear model, you can see that the treatment effect should be added to (or subtracted from) the error effect if the model is to hold true.

3.   Analysis of variance is a parametric test, that is, it is tied to a population. Therefore, certain assumptions must be made about the population from which the sample of measurements is drawn. Specifically, it is assumed that the variable in question is normally distributed in the population. It is generally agreed, however, that reasonable departures from normality do not seriously affect the validity of the results.

8.4 EVALUATION OF DATA

Back in Chapter 1, we discussed some of the various ways in which statistics can be misused. One of the problems mentioned at the time was the "garbage in–garbage out" syndrome which seems to affect a significant number of research projects. For example, if the experimental data do not begin to conform to the major assumption of the test we are using, then the results are certainly questionable, and the sophistication of test, calculator, or computer can do little to help. It is therefore incumbent upon the investigator to examine his or her data *prior* to performing an analysis, instead of blindly proceeding with the analysis.

As we have seen, it is especially important to examine the treatment-group variances for heterogeneity.

If only two treatments are involved, the question of homogeneity of variances can be easily answered by setting up a variance ratio with the larger variance in the numerator and the smaller variance in the denominator. This yields an *F*-value that can be evaluated by looking in

Table VII. In this situation the numerator degrees of freedom is $(N_A - 1)$ and the denominator degrees of freedom is $(N_B - 1)$, or the other way around if the variance of group B is larger than that of A. If the F-value is significant, we can then conclude that the variances are significantly different and therefore not homogeneous.

If three or more treatments are involved, then the simple variance ratio is not applicable. With three or more group variances, it is necessary to perform *Bartlett's test* for homogeneity of variances. Bartlett's test is not difficult but it is rather tedious and should be done by a computer if one is available. If not, a step-by-step description of Bartlett's test can be found in Sokal and Rholf, one of the references listed in the bibliography. Sokal and Rholf also offer an explanation of the *F-max* test, which is a simpler but possibly less reliable test of homogeneity of variances.

Although it continues to be commonly used, in statistical circles there is currently some doubt expressed about the value of Bartlett's test. For one thing, statisticians point out that analysis of variance is a *robust* test, which means that its reliability tends to suffer relatively little from reasonable departures from its assumptions. In other words, if the group variances are not homogeneous (but not wildly different) the F-value is still reliable, especially if it is not at borderline significance.

If the investigator obtains experimental data that clearly violate the assumptions of analysis of variance beyond a reasonable point, then the application of a *transformation* may or may not be helpful. The use of transformations will be discussed in the next section. Finally, if all else fails, the researcher may resort to a *nonparametric* or so-called "distribution-free test." Examples of nonparametric tests will be presented in Chapter 12.

8.5 TRANSFORMATIONS

When the data appear to radically depart from the assumptions of a statistical test, we can sometimes save the day by using a *transformation*. The application of transformations to data should not be thought of as a "fudge factor" but simply as a change in the scale of measurement that may or may not help the data to meet the assumptions of a test. For a more detailed discussion of transformations and their applications, the reader is referred to the reference by Steel and Torrie that is listed in the bibliography. In this section we will present two of the more common transformations as illustrations of how the scale of measurement can be changed.

The Logarithmic Transformation

In Section 8.4 we said that, in analysis of variance, the treatment effects must be additive if the linear model is to be satisfied. If the treatment effects are multiplicative, gross differences in the treatment variances will be produced. This may be better understood from the following illustration.

Suppose we have a distribution such as:

$$2$$
$$3$$
$$1 \qquad n = 5$$
$$5 \qquad \overline{X} = 3$$
$$4 \qquad \sum(X - \overline{X})^2 = 10$$

The mean of this distribution is 3. Now, if we sum the squared deviations of all members of the distribution from the mean, we obtain an SS of 10. This, divided by $(n - 1)$, or 4, yields a variance of 2.5.

Now, let's add a constant, say 3, to each member of the original distribution. This results in a new distribution,

$$2 + 3 = 5$$
$$3 + 3 = 6$$
$$1 + 3 = 4 \qquad n = 5$$
$$5 + 3 = 8 \qquad \overline{X} = 6$$
$$4 + 3 = 7 \qquad \sum(X - \overline{X})^2 = 10$$

which has a mean of 6. Again, we determine the SS, and obtain 10, which, divided by 4, yields a variance of 2.5. Thus it may be seen that adding a constant to a distribution produces a new distribution with a greater mean but with the *same variance.*

If we now multiply each number of the original distribution by 3 we will obtain a new distribution,

$$2 \times 3 = 6$$
$$3 \times 3 = 9$$
$$1 \times 3 = 3 \qquad n = 5$$
$$5 \times 3 = 15 \qquad \overline{X} = 9$$
$$4 \times 3 = 12 \qquad \sum(X - X)^2 = 90,$$

which has a mean of 9. This time, when we compute the SS, we obtain a total of 90. Dividing this by 4, we obtain a variance of 22.5. The variance of the new distribution is considerably different from the variance of the original distribution, and the mechanics of analysis of variance cannot be legitimately applied. Therefore, if the treatment effects are multiplicative rather than additive, it is not appropriate to apply the analysis of variance, at least to the raw data.

In a situation where the treatments show multiplicative effects, we may be able to appropriately change the scale of measurement by using a *log transformation*. For example, given the following two groups,

A	B
40	80
20	40
30	60
15	30
25	50

we can see that the treatment effect is multiplicative, since, in B, each variate of A is multiplied by a factor of 2. Now, since the log xy is equivalent to log x + log y, converting the raw data to equivalent logarithms will change multiplicative effects to additive effects, as shown in Table 8.3.

TABLE 8.3

A	$(X - \bar{X}_A)$	$(X - \bar{X}_A)^2$	B	$(X - \bar{X}_B)$	$(X - \bar{X}_B)^2$
1.602	0.212	0.044944	1.903	0.212	0.044944
1.301	0.089	0.007921	1.602	0.089	0.007921
1.477	0.087	0.007569	1.778	0.087	0.007569
1.176	0.214	0.045796	1.477	0.214	0.045796
1.398	0.008	0.000064	1.699	0.008	0.000064
6.954		0.106294	8.459		0.106294
$\bar{X}_A = 1.390$			$\bar{X}_B = 1.691$		

Computing the variance with the sums of squares obtained in the table, we divide the SS in each case by $(n - 1)$, or 4, and obtain

$$S_A^2 = \frac{\Sigma(X - \bar{X}_A)^2}{n_A - 1} = \frac{0.106294}{4} = 0.026573,$$

and

$$S_B^2 = \frac{\Sigma(X - \bar{X}_B)^2}{n_B - 1} = \frac{0.106294}{4} = 0.026573.$$

Thus the variances have been made equal by the log transformation, and we may proceed with the analysis of variance, using the transformed data. Since in our example we worked with contrived data, we obtained textbook results, but with actual experimental data we would be satisfied if we obtained variances showing an acceptable degree of homogeneity.

The Arcsin Transformation

Biological data may occasionally take the form of percentages. In general, if the data tend to range between 30 and 80 percent, we can probably assume that the population distribution is normal because the *mean* more than likely is somewhere around 50 percent. In this case, therefore, a transformation would not be necessary. Suppose, however, that our data suggest that the population mean is somewhere in the neighborhood of 90 percent. In this situation the curve is likely to be severely skewed because we can go only 10 percentage points to the *right* of the mean, but could go 90 percentage points to the left of the mean. Quite obviously, therefore, the data are not normally distributed in the population, and we have a problem meeting one of the assumptions of analysis of variance. In this situation we may solve the problem relatively simply by applying the *arcsin transformation*.

For example, a value such as 74 percent is first converted to the decimal form, 0.74. We then find the square root of that decimal, or

$$\sqrt{0.74} = 0.86.$$

We now look in a table of natural sines for an angle, the sine of which is 0.86. This turns out to be slightly more than $59°20'$. Expressing $20'$ as a decimal, we arrive at $59.33°$ as the arcsin transformation of 74 percent.

While this procedure is not difficult, it could be quite tedious when dealing with large quantities of data. Happily, the mathematicians have once more come to our rescue by providing Table VIII, which may be used to find the transformed values directly.

The following example will illustrate the arcsin transformation:

Example. An experiment was performed to determine significant differences, if any, among the effects of three inhibitors, A, B, and C, on cellular uptake of epinephrine labeled with C^{14}. The data in Table 8.4 are expressed in percent uptake per milligram dry weight of cells. Each treatment was replicated eight times.

Each of the percent values in the original data was first converted to the decimal form, and the decimal was then transformed to degrees by the use of Table VIII. Having transformed the data, we may now perform analysis of variance in the usual manner.

TABLE 8.4

A		B		C	
%	*Degrees*	%	*Degrees*	%	*Degrees*
65	53.73	68	55.55	70	56.79
72	58.05	67	54.94	74	59.34
59	50.18	70	56.79	60	50.77
69	56.17	66	54.33	74	59.34
70	56.79	63	52.53	71	57.42
68	55.55	57	49.02	69	56.17
70	56.79	71	57.42	66	54.33
73	58.69	68	55.55	71	57.42

8.6 ANALYSIS OF A COMPLETELY RANDOMIZED DESIGN

In Section 8.2, our method of calculating the *F*-value was intended as an aid to understanding the basic principles of analysis of variance. In a practical situation the data would almost certainly be analyzed either by computer or by using a series of formulas designed for a desk calculator. At this point we are ready to use the "machine method" to analyze data derived from a practical experimental situation.

Example.* Fascioliasis is a widespread parasitic disease that is produced by *Fasciola hepatica*, which is a flatworm commonly called the liver fluke. The adult worms live in the bile duct of the host. Investigators (Isseroff, *et al.*) have found that the adult worms secrete significant amounts of certain amino acids, especially proline. It has also been established that the host characteristically exhibits an anemia that involves a reduction in circulating red blood cells.

An experiment was designed to test the following questions:

1. Is the reduction in the number of the host's red blood cells associated with the secretion of amino acids by the worm?

2. Is the secretion of significant amounts of proline specifically associated with the reduction in the number of red blood cells?

Forty noninfected Wistar rats were selected, keeping such factors as age and weight as constant as possible. They were assigned randomly, ten each, to four different treatments. An apparatus was adapted (T. Sawma)

---

* This example is adapted from experiments performed by T. Sawma and D. Reno as part of a research program directed by Dr. H. Isseroff.

so that materials could be infused through a cannula sirectly into the bile ducts of the rats. The treatment groups are described as follows:

1. The rats in group D were infused with 20-millimolar proline in physiological saline solution.

2. The rats in group E were infused with a smaller concentration (2-millimolar) proline in physiological saline solution.

3. The rats in group F were infused with a "cocktail" consisting of seven amino acids (excluding proline) that have been shown to be excreted by the worm. This infusion also involved physiological saline.

4. To account for the possible independent variable that consisted of the presence of the cannula, the pressure associated with the infusion process, and the surgical procedures involved, a fourth group (C) was established as a control group. The rats in this group were infused with a physiological saline solution only.

Table 8.5 shows the data derived from RBC counts on the individual subjects, reported in millions per cubic millimeter and rounded to the hundredth place. As before, the group statistics have subscripts that cor-

TABLE 8.5

D	E	F	C
20 mmol. proline	*2 mmol. proline*	*Amino-acid mixture*	*Saline only*
5.61	6.07	5.69	7.35
5.40	5.20	5.54	7.11
5.26	5.69	5.35	6.99
4.99	5.43	5.11	6.72
5.44	5.87	5.94	7.16
5.13	5.55	5.25	6.85
5.21	5.64	6.02	6.94
5.52	5.95	5.64	7.25
4.79	5.20	5.11	6.51
4.92	5.40	5.04	6.65
$\sum X_D = 52.27$	$\sum X_E = 56.00$	$\sum X_F = 54.69$	$\sum X_C = 69.53$
$S_D^2 = 0.073$	$S_E^2 = 0.091$	$S_F^2 = 0.124$	$S_C^2 = 0.074$
$\bar{X}_D = 5.23$	$\bar{X}_E = 5.60$	$\bar{X}_F = 5.47$	$\bar{X}_C = 6.95$
$n_D = 10$	$n_E = 10$	$n_F = 10$	$n_C = 10$

$$\sum X^2 = 1372.61 \qquad\qquad \sum X = 232.49$$

respond to the letter (D, E, F, or C) assigned to a specific group. The total sample is $n_T = 40$. Note that we have computed the sum of each treatment group because we will need these figures in our calculations. Also, there are four treatments, so $k = 4$.

The experimental design presented here is called a *completely randomized design* and is analyzed by *single-classification* or *one-way* analysis of variance. Actually, this is the same type of analysis that we carried out in Section 8.2, but in this section we will use the more standard "machine method."

At this point we will begin a step-by-step analysis of the data in Table 8.5. Be sure to refer to Table 8.5 as we go along:

1. First, the null hypothesis that we are testing is

    $$\mu_D = \mu_E = \mu_F = \mu_C,$$

    which is a symbolic way of stating that no statistically significant differences exist among the treatments. To put it another way, we are saying that all 40 rats in our sample belong to the same statistical population as far as red-blood-cell counts are concerned.

2. An examination of the treatment error (within group) variances shows that they are homogeneous within satisfactory limits. If we were to find a suspicious degree of heterogeneity among the treatment-group variances, we would perform Bartlett's test or the F-max test. If we found an unreasonable departure from homogeneity, we could try a transformation, or proceed with the Kruskal–Wallis test, which is described in Chapter 12.

3. Next, we compute the sum of squares (SS) for the total distribution of 40 measurements. Once again, remember that the "sum of squares" is a shorthand version of *sum of squared deviations from the mean*. We therefore obtain the sum of squared deviations of all 40 measurements from the grand mean of the 40 measurements. Using a desk calculator, we obtain $\sum X^2$ and the $\sum X$ by $(5.61^2 + 5.40^2 + 5.26^2 + \cdots + 6.51^2 + 6.65^2)$. Most desk calculators will yield $\sum X^2$ and $\sum X$ with one run through the data. In our case, we find $\sum X^2 = 1372.61$ and $\sum X = 232.49$. We now put these values into the machine formula for the sum of squares as follows:

    $$SS_{Total} = \sum X^2 - \frac{(\sum X)^2}{n_T}$$

    $$= 1372.61 - \frac{(232.49)^2}{40}$$

    $$= 1372.61 - 1351.29 = 21.32.$$

4. Next, we compute the sum of squares for treatment effects (among groups), using a machine formula in which we use the *sums* of the measurements in the individual treatment groups.

$$\text{SS}_{\text{Treatments}} = \frac{(\sum X_{\text{D}})^2}{n_{\text{D}}} + \frac{(\sum X_{\text{E}})^2}{n_{\text{E}}} + \cdots + \frac{(\sum X_{\text{C}})^2}{n_{\text{C}}} - \frac{(\sum X)^2}{n_{\text{T}}}$$

$$= \frac{(52.27)^2 + (56.00)^2 + \cdots + (65.53)^2}{10} - \frac{(232.49)^2}{40}$$

$$= 1369.36 - 1351.29 = 18.07.$$

Note that the quantity $(\sum X)^2/n_{\text{T}}$ is used in computing both the total and the treatment sums of squares. This quantity is sometimes called the "correction factor."

5. Since the "total" sum of squares consists of the sum of squares for treatment *plus* the sum of squares for error (within groups), we can use the following shortcut to compute the sum of squares for error:

$$\text{SS}_{\text{Error}} = \text{SS}_{\text{Total}} - \text{SS}_{\text{Treatment}}$$

$$= 21.32 - 18.07 = 3.25.$$

6. Having computed the sums of squares for "total," "treatments," and "error," we now set up an analysis-of-variance table, which is shown below as Table 8.6.

TABLE 8.6

Source	SS	d.f.	MS
Treatment	18.07	3 $(k - 1)$	6.02
Error	3.25	36	0.090
Total	21.32	39 $(n_{\text{T}} - 1)$	

Looking at Table 8.6, note that the treatment degrees of freedom is $(k - 1)$, which in this case is three. The degrees of freedom for "total" is $n_{\text{T}} - 1$, or 39. Filling in the degrees of freedom for error is very simple because the degrees of freedom for treatment and for error must add to the degrees of freedom for total. Our error degrees of freedom is therefore 36. The *mean square* (MS), which is another term for a variance *estimate*, is computed by dividing the SS by the degrees of freedom. This yields the mean squares (variance estimates) associated with the variability among the treatment means and the variability *within* groups. Dividing the MS for treatments by the MS for error, we obtain an *F*-value which we look up in Table VII with 3 and

36 degrees of freedom:

$$F_{[3,\,36]} = \frac{MS_{Treatments}}{MS_{Error}}$$

$$= \frac{6.02}{0.090} = 66.89\ (P < 0.001).$$

We find that the obtained F-value (66.89) is significant well beyond the 0.001 level. The symbol ($P < 0.001$) indicates that the probability of committing a Type I error if we reject the null hypothesis is less than 0.001. If it happens to be a dull evening with nothing but reruns on television, you might try analyzing the data in Table 8.5 by the method described in Section 8.2. You will find a slight difference in the F-value because of differences in rounding, but you will see that the "machine method" yields essentially the same result as the more basic "principles" approach.

7.   Since the F-value obtained from our data is significant beyond the 0.001 level, we should have no difficulty deciding to reject the null hypothesis. On the basis of this decision, we can therefore state that we have experimental and statistical evidence that supports our contention that statistically significant differences exist among the treatment means.

The analysis is still not finished because now another problem arises. Are all of the treatment means significantly different from one another? If not, which means are significantly different from which? Our highly significant F-value certainly suggests that something is going on, but we will not be in a position to draw conclusions from our data until we find what exactly *is* going on that has produced that significant F-value.

Having obtained a significant F-value, our next step is to carry out a test by which we can compare all possible pairs of means, which in this case is six *pairs*. This kind of test is called an *a posteriori* test of means and is applied only when a significant F-value is obtained. There are a number of such tests available, each with its individual advantages and disadvantages. The reader is referred to the text by Steel and Torrie for a more complete description of *a posteriori* tests, since to examine all of them in detail is beyond the scope of this book. For our purposes we will use the Student–Neumann–Kuels (SNK) test, which is generally regarded as a reliable *multiple-range* test.

8.7 THE STUDENT–NEUMANN–KUELS TEST

Multiple-range tests are based on the general concept that, as the number of treatment groups increases, the greater is the likelihood that the extremes or near extremes will show false significant differences when they are compared by using a *t*-test. In other words, if we draw enough treatment groups from a population, the extremes may well show significant differences even though they actually belong to the same distribution of means. We therefore need to evaluate the differences between extremes and near extremes with a test that has a built-in correction factor that takes into consideration the number of means (k) being compared.

The Student–Neumann–Kuels, or SNK test, provides a kind of sliding scale of minimum differences that must be equaled or exceeded in order to establish significance. In other words, if we have five treatment means that range in increasing value from $\overline{X}_1$ to $\overline{X}_5$, there must be a greater minimum difference between $\overline{X}_5$ and $\overline{X}_1$ than between $\overline{X}_4$ and $\overline{X}_3$ or between $\overline{X}_3$ and $\overline{X}_1$. This should become more clear as we go through the following step-by-step SNK procedure, using our experimental data from Tables 8.5 and 8.6.

1. In our example, we obtained a highly significant *F*-value (66.89, $P < 0.001$), which indicates that statistical significance exists between at least one pair of treatment means and possibly more. Having obtained a significant *F*-value from the general analysis, we can now compare the various mean pairs to see which are statistically different from which.

2. First, we calculate an estimate of the standard error of the mean as follows, using the error mean square from Table 8.6 as the estimate of σ^2. The number of replications per treatment is substituted for n. Thus,

$$S_{\overline{X}} = \sqrt{\frac{S^2}{n}} = \sqrt{\frac{0.090}{10}} = 0.09487.$$

3. We now turn to Table XIII in the appendix. In using the table, we go down the left side to find the error degrees of freedom associated with the analysis. From Table 8.6 it may be seen that the error degrees of freedom is 36, but in Table XIII there is a gap between 30 and 40. While it may be argued that 36 is closer to 40, we should use the next *lowest* figure (30) because this makes it more difficult, not easier, to find significance between any given pair of means. (An even better solution, obviously, is to look up the actual value in a more extensive table.) Now find 30 degrees of freedom and move across the table to the right. Note that the *q*-value increases as the number of treatment means (k) increases. We shall see that this is the built-in correction factor that was mentioned earlier.

4. Next, we set up Table 8.7. The top row of the table contains the *k*-values relevant to our experiment. For example, $k = 2$ is associated with any two means that are adjacent; $k = 3$ refers to the comparison of two means that are at the extremes of three means, and $k = 4$ is

used to determine the minimum difference that must be found between two means that are the extremes of four means. From Table XIII we now obtain the q-values associated with 30 degrees of freedom and the k-values 2, 3, and 4, respectively. These are entered in the second row of Table 8.7.

TABLE 8.7

k	2	3	4
q	2.89	3.49	3.85
$qS_{\bar{X}}$	0.28	0.33	0.37

Finally, we multiply each q-value by $S_{\bar{X}}$, which we have already computed as 0.09487. This gives us $qS_{\bar{X}}$, and each of these values, entered in the bottom row of Table 8.7, is the minimum difference that must be found between two means associated with a specific value of k if the two means are to be considered statistically significantly different.

5. The next step is to arrange the four treatment means in an array, ranging from the smallest to the largest, as shown in Table 8.8.

TABLE 8.8

20 mmol. pro-line	Amino-acid mixture	2 mmol. pro-line	Saline only
$\bar{X}_1$	$\bar{X}_2$	$\bar{X}_3$	$\bar{X}_4$
5.23	5.47	5.60	6.95
D	F	E	C

6. We are now ready to compare all six possible pairs of means, beginning with $\bar{X}_1$ and $\bar{X}_4$, as shown below:

$$\bar{X}_4 - \bar{X}_1 = 1.72 > 0.37*$$

$$\bar{X}_3 - \bar{X}_1 = 0.37 > 0.33*$$

$$\bar{X}_2 - \bar{X}_1 = 0.24 < 0.28 \quad \text{N.S.}$$

$$\bar{X}_3 - \bar{X}_2 = 0.13 < 0.28 \quad \text{N.S.}$$

$$\bar{X}_4 - \bar{X}_2 = 1.48 < 0.33*$$

$$\bar{X}_4 - \bar{X}_3 = 1.35 > 0.28*$$

Note above that the differences between the various pairs of means are each compared to the corresponding minimum difference found in Table 8.7. For example, $\overline{X}_3$ and $\overline{X}_1$ are the extremes of three means ($k = 3$); therefore the difference between $\overline{X}_3$ and $\overline{X}_1$ must equal or exceed 0.33. Since the actual difference between $\overline{X}_3$ and $\overline{X}_1$ is 0.37, we note that it is significant. Also, because $\overline{X}_3$ and $\overline{X}_2$ are the "extremes" of only two, or adjacent means, k is therefore 2 and the difference between them must equal or exceed 0.28. The actual difference between $\overline{X}_3$ and $\overline{X}_2$ is only 0.13 and is therefore nonsignificant.

7. If the above results are translated into terms that are relevant to our experiment, it may be seen that the control group (saline only) differs significantly from the other three treatments. In other words, our results strongly suggest that proline and other amino acids secreted by the worm are somehow associated with the reduction in the red cell numbers that is typically observed in fascioliasis. The differences between the effects of proline and the other amino acids are not conclusive; however, the slight significance found between 2 mmol. proline and 20 mmol. proline suggests the possibility that proline may have an especially strong association with the reduction in number of circulating red cells.

Looking back at our experiment, it might be suggested that more definitive results could be obtained if the power of the test were increased by using larger numbers of replications per treatment. You will recall that, if the power of a test increased, the likelihood of detecting small but interesting significant differences is also increased. Increasing the treatment n-numbers decreases the value of $S_{\overline{X}}$ because it is calculated by

$$S_{\overline{X}} = \sqrt{\frac{S^2}{n}};$$

and, since the minimum difference between a given pair of means is determined by $q S_{\overline{X}}$, it is obvious that a smaller $S_{\overline{X}}$ would increase the chances of detecting small differences. Once again, however, we should recall the diminishing returns from an increase in n, and the problems of cost and time are, as always, an important part of the planning process.

8.8 UNEQUAL TREATMENT GROUPS

Up to now, we have been working with examples that had equal numbers of replications per treatment. As a practical aspect of experimental design, it is always a good idea to keep treatment groups equal; indeed, this should be one of the things kept uppermost in mind when planning an experiment.

There is, however, a basic law of research, and it goes like this: *If anything can go wrong, it will, and things that cannot go wrong probably also will.* The neat and tidy experiments found in books like the one you are reading tell little or nothing of the frustrations of the researcher who must contend with critical power failures, faulty equipment, uncooperative weather, or rats that suddenly die at a critical point, from no discernible cause other than a desire to get even with the experimenter.

There are, therefore, many situations that arise in biological research in which we have no choice but to deal with unequal treatment groups. This is done by using the same formulas for the sums of squares used in the fascioliasis example, except that when computing the treatment SS, *the square of each treatment sum is divided by the number of measurements that make up the sum.* For example, if we have the following treatments,

A	B	C
$n_A = 8$	$n_B = 10$	$n_C = 12$
$\sum X_A = 15$	$\sum X_B = 18$	$\sum X_C = 22$

we calculate the SS for treatments as follows:

$$SS_{Treatments} = \frac{(\sum X_A)^2}{n_A} + \frac{(\sum X_B)^2}{n_B} + \frac{(\sum X_C)^2}{n_C} - \frac{(\sum X)^2}{n_T}$$

$$= \frac{(15)^2}{8} + \frac{(18)^2}{10} + \frac{(22)^2}{12} - \frac{(55)^2}{30}.$$

Meanwhile, the SS for "total" and the SS for "error" are calculated in the same way as they were in the fascioliasis example.

Unfortunately, the SNK test becomes more complicated when the design has unequal treatment n-numbers. You should recall that $S_{\bar{X}}$ is calculated as:

$$S_{\bar{X}} = \sqrt{\frac{S^2}{n}},$$

which translates to

$$S_{\bar{X}} = \sqrt{\frac{S_1^2}{n_1} + \frac{S_2^2}{n_2} + \cdots},$$

so that the formula for $S_{\bar{X}}$ used in the SNK test holds true only if n_1, n_2, n_3, etc., are all equal. A version of the SNK test that can be done with unequal treatment numbers is available, however, and the reader is referred to Sokol and Rholf for a good step-by-step example of how it is done. If the treatment n-numbers vary only slightly, another possibility is to perform the SNK test as usual with the *smallest* n-number used as n in the formula for $S_{\bar{X}}$.

8.9 MODEL I AND MODEL II ANALYSIS OF VARIANCE

So far we have discussed examples of Model I analysis of variance only. An analysis of variance design is considered a Model I if the treatments are "fixed"; that is, if they are selected by the experimenter for a particular reason and he or she is interested only in any differences that may exist among the specific treatments selected. On the other hand, a Model II analysis of variance involves *randomly* selected treatments; that is, the treatments are selected in a random fashion from a *population* of treatments.

To use another silly example, suppose that we wish to compare three headache remedies such as aspirin, Anacin, and Excedrin to see whether there are significant differences in their effectiveness. In this situation we are interested only in these three drugs; the treatments are therefore fixed and we are working with a Model I analysis of variance. To put it differently, we want to see if there would be statistically significant differences among the three *fixed* theoretical *populations* of headache sufferers who *could* be given these drugs.

On the other hand, if we were interested in finding out whether there is a significant variability among headache remedies, we would then randomly select three, four, five, or whatever number of headache remedies is feasible from the total *population of headache remedies*, which must include thousands of different brands. In this case we are not interested in differences among specific or "fixed" treatments; we are instead interested in the degree to which the effectiveness of headache remedies in general may vary.

Space limitations do not permit us to go into the details of Model II analysis of variance. In-depth descriptions of Model II analyses can be found in references listed in the bibliography, particularly in Snedecor, Steel and Torrie, and Sokol and Rholf. Woolf is highly recommended for an excellent and understandable discussion of the Model II design. It is probably correct, however, to say that the vast majority of biological research problems involve the Model I design. In the next chapter, on the other hand, we will see that a useful procedure called a "nested" design does contain elements of the Model II.

8.10 SUMMARY

In this chapter you have been introduced to the basic principles and assumptions that pertain to analysis of variance. It is hoped that you have at least gained an understanding of the basic principles and precautions that should be .observed when using analysis of variance and thereby avoid becoming another victim of the "garbage in–garbage out" syndrome. You should now be more aware of the need to carefully evaluate data that are put into a computer program, and you should also be able to more intelligently interpret the computer printout.

PROBLEMS

8.1 The data in Table 8.9 represents the height (in centimeters) of plants grown in three different media with five replications per medium. Analyze the data to determine whether statistically significant differences exist among the treatment means. If the F-value is significant, test for statistically significant difference between all pairs of treatment means (see Sections 8.6 and 8.7).

TABLE 8.9

A	B	C
10	16	15
14	18	12
18	22	8
15	18	10
12	15	13

8.2 Eight samples were taken from each of three locations in a stream to determine whether a statistically significant difference in total nitrogen content existed among the locations. The data in Table 8.10 are expressed as milligrams/100 grams. Analyze the data for statistical significance, and if a significant F-value is obtained, then test for statistically significant differences between all pairs of location means (see Sections 8.6 and 8.7).

TABLE 8.10

Location		
A	B	C
222	326	263
300	275	260
262	218	299
264	207	221
200	272	198
211	268	211
267	308	266
326	229	319
$\overline{X}_A = 256.50$	$\overline{X}_B = 262.88$	$\overline{X}_C = 254.63$

8.3 Four strains of rats were selectively bred for differences in blood pressure in order to determine the possible effect of heredity on blood pressure. A, B, C, and D are given in Table 8.11.

TABLE 8.11

| | | Strain | |
A	B	C	D
84	87	89	89
82	84	94	86
86	84	92	88
89	92	91	93
85	88	92	85
85	89	91	85
92	92	95	89
80	89	89	90
79	87	87	90
83	88	91	93
$\overline{X}_A = 84.50$	$\overline{X}_B = 88.00$	$\overline{X}_C = 91.10$	$\overline{X}_D = 88.8$

Analyze the data for statistical differences among the means. If a statistically significant F-value is found, then test for statistically significant differences among all possible mean pairs. (See Sections 8.6 and 8.7.)

8.4 A sample was collected from each of three geographically isolated populations of a certain species of bird. The bill length of each specimen was measured to the nearest tenth of a millimeter, yielding the data in Table 8.12. Analyze the data for statistically significant differences among the populations (see Section 8.8).

TABLE 8.12

| | Population | |
A	B	C
4.2	3.8	3.0
3.3	4.1	3.5
2.8	5.0	4.5
4.3	4.6	4.4
3.7	5.1	
4.5		
3.6		

8.5 Two different baits were tested for significant difference in terms of consumption by wild rats. The data in Table 8.13 were obtained from five locations per bait, and are expressed in percentage consumption. Analyze these data for significance, applying the arcsin transformation for percentage data (see Section 8.5).

TABLE 8.13

Bait	
A	B
10	15
15	20
12	16
20	25
14	20

8.6 Examination of the three sets of treatment data in Table 8.14 reveals the presence of multiplicative effects. Compute the mean and variance of each set of raw data; then apply a log transformation and again compute the mean and variance of each set. Compare the variances with respect to homogeneity (see Section 8.5).

TABLE 8.14

I	II	III
12	16	31
15	29	26
9	23	20
11	22	36
13	23	28

Analysis of Variance— Part II

9.1 INTRODUCTION

In Chapter 8 we discussed the fundamentals of analysis of variance, and we also considered the "one-way" analysis of a *completely randomized* design. In this chapter we will introduce some of the more sophisticated but commonly used experimental designs and their analysis. Actually, these designs are relatively simple and the procedures are quite routine once you grasp the general patterns.

9.2 THE NESTED DESIGN

The *nested*, or *hierarchical* design is very useful in a number of biological research situations. Basically, the nested design is an extension of the one-way analysis of variance with which we are already familiar, except that it consists of several one-way analyses, each nested within a higher "tier." Perhaps the fastest and easiest way to explain this type of design is to work through the following example:

Example*. Suppose that we have an approximately one-quarter square-mile area of a lake that is used as a dump site for dredgings from a nearby harbor. We are interested in whether or not differences in certain chemical and biotic factors exist between the dump area and other parts of the lake where no dumping has taken place. Or, to put it in more practical terms, what effects, if any, do the harbor dredgings have on the lake ecosystem? We therefore select a control area that matches the dump area as closely as possible in size, depth, seasonal changes, etc., except that it is "uncontaminated" by harbor dredgings.

---

* Adapted from a research project performed under the direction of Dr. R. A. Sweeney.

At this point we could randomly select a sampling site in each of the two areas. We could then draw n samples, or replicated measurements, from each of the two sampling sites and test the resulting data for significance with a t-test. Unfortunately, this procedure would pose a troublesome question. How do we know that our randomly selected sampling sites adequately reflect the actual conditions that exist in two relatively large areas of the lake? For example, if we again go out and randomly select different sampling sites in the two areas, is it possible that our results would be very different? Which pair of sites would best reflect the general differences, if any, that exist between the two areas?

To account for these possibilities, we will randomly select *several* sampling sites in each of the two areas. To simplify things, we will use only three sampling sites per area, but in practice a larger number of sites would probably be more appropriate (Fig. 9.1).

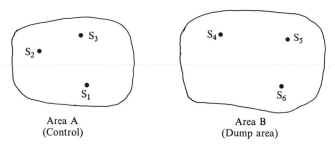

Area A
(Control)

Area B
(Dump area)

Fig. 9.1 Selection of sampling sites in two isolated areas.

After the sampling sites have been randomly selected, we now draw three replicate samples from each site and measure one of the parameters of interest, such as the concentration of a certain chemical. This yields the data for our experiment, organized as shown in Table 9.1.

Looking at Table 9.1, we can see that the sampling sites are "nested within" the areas. We have the measurements taken at each site, the totals for the sites, and the totals for the areas. We can now test for two things: (1) the difference between Areas A and B, and (2) the variability of the sampling sites within areas. If we fail to find a significant variability among the sampling sites *within areas*, then a significant difference between areas would suggest that the dredgings are having effects on area B. In other words, the variability is due to differences between the areas and not to variability among the sampling sites. In this kind of situation, however, it is highly likely that we *will* find variability among the sampling sites. Even if it should be significant, however, we can still test to see whether the difference between the areas is significantly larger than the variability

TABLE 9.1

	Area A (Control)			Area B (Dump area)	
S_1	S_2	S_3	S_4	S_5	S_6
18	19	18	21	19	19
16	20	18	20	20	23
16	19	20	18	21	21
50	58	56	59	60	63

$$\text{Total}_A = 164 \qquad\qquad \text{Total}_B = 182$$

$$\text{Grand total} = 346$$

among the sampling sites within areas. All of this may be confusing at first, but if you review what we have said so far and think about it carefully, you will see that the whole idea is quite simple and logical.

Our next step is to compute the appropriate sums of squares as follows:

1. First, we compute the sum of squares for "total" in the usual way by:

$$SS_{\text{Total}} = \sum X^2 - \frac{(\sum X)^2}{n_T}$$

$$= 18^2 + 16^2 + 16^2 + \cdots + 23^2 + 21^2 - \frac{(346)^2}{18}$$

$$= 6704 - 6650.89 = 53.11.$$

2. Next, we compute the sum of squares for "areas" as it would be done in any one-way analysis of variance. Table 9.1 shows the totals for areas as 164 and 182. The "correction factor" is $(\sum X)^2/n_T$, or 6650.89. Thus,

$$SS_{\text{Areas}} = \frac{(\sum X_A)^2}{n_A} + \frac{(\sum X_B)^2}{n_B} - \frac{(\sum X)^2}{n_T}$$

$$= \frac{(164)^2}{9} = \frac{(182)^2}{9} - 6650.89$$

$$= 6668.89 - 6650.89 = 18.$$

3. Our next step is to compute the sum of squares for sampling sites. This involves a slight complication because there are two factors that contribute to the differences among the sites. First, there may be (and probably are) differences from site to site because of spatial and localized conditions found at each site. In fact, you will recall that this is the major reason why we set up this design in the first place. Secondly, the sites may differ because of *a difference between the areas in which they are nested.* Therefore, to compute the "pure" sum of squares for sampling sites, we must subtract the squared totals for areas, thus removing that portion of the variability among sites that is due to and originating from any difference that exists between the areas. This yields the sum of squares for "sites within areas" as follows:

$$SS_{\text{Sites within areas}} = \frac{(\sum S_1)^2 + (\sum S_2)^2 + (\sum S_3)^2 + \cdots + (\sum S_6)^2}{n_{\text{Site}}}$$

$$- \frac{(\sum X_A)^2 + (\sum X_B)^2}{n_{\text{Area}}}$$

$$= \frac{(50)^2 + (58)^2 + (56)^2 + \cdots + (63)^2}{3} - \frac{(164)^2 + (182)^2}{9}$$

$$= 6683.33 - 6668.89 = 14.44.$$

In this computation you should note the general procedure of dividing by the number of measurements that make up a given total. For example, the site totals each contain three values and the area totals each contain nine measurements.

4. Finally, we need to calculate the sum of squares for error. Since the SS for error plus the sums of squares found in steps (2) and (3) must add to the total SS, the error sum of squares can be computed simply as follows:

$$SS_{\text{Error}} = SS_{\text{Total}} - (SS_{\text{Areas}} + SS_{\text{Sites within areas}})$$

$$= 53.11 - (18 + 14.44) = 20.67.$$

Having computed the necessary sums of squares, we can now set up an analysis-of-variance table as in Table 9.2.

TABLE 9.2

Source	SS	d.f.	MS	Expected MS
Areas	18.00	1	18.00	σ^2 + Sites w/areas + Areas
Sites within areas	14.44	4	3.61	σ^2 + Sites w/areas
Error	20.67	12	1.72	σ^2
Total	53.11	17		

Looking at Table 9.2 we can see that the degrees of freedom for areas is 1, which is consistent with the $(k - 1)$ formula for "treatments" that we used in Chapter 8. The degrees of freedom for "sites within areas" is 4; i.e., within each area there are three sites with "sites -1" degrees of freedom, or 2. The total is therefore 4. As usual, the degrees of freedom for "total" is $n_T - 1$, which is $18 - 1$, or 17. The error degrees of freedom is therefore 12, since the various degrees of freedom values must add to 17.

The mean squares for "areas," "sites within areas," and "error" are all computed in the usual way by dividing each SS by the associated degrees of freedom. Table 9.2 also shows the "expected mean squares" associated with each source of variation. Each expected MS consists of the variance estimates that are contained within it. For example, the MS for "areas" contains an estimate of σ^2 *plus* variability due to differences among the sampling sites *plus* that part of the variance that is due to any difference that may exist between the areas. The MS for "sites within areas" contains an estimate of σ^2 *plus* an estimate of the variance among all the sites that could conceivably be randomly selected within the areas. If you review our brief discussion of Model I and Model II designs in Chapter 8, you can see that "sites within areas" is a random variable, while "areas" is a fixed variable. Looking again at the expected mean squares, it may be seen that each "tier" must be tested over the one immediately below. Thus,

$$F_{[1,4]} = \frac{MS_{\text{Areas}}}{MS_{\text{Sites within areas}}}$$

$$= \frac{18.00}{3.61} = 4.99$$

and

$$F_{[4,12]} = \frac{MS_{\text{Sites within areas}}}{MS_{\text{Error}}}$$

$$= \frac{3.61}{1.72} = 2.10.$$

Turning to Table VII, we find that the F-value for "areas" (1 and 4 d.f.) is not statistically significant, and that the F-value for "sites within areas" (4 and 16 d.f.) is also not significant.

From our experiment, we therefore have no evidence that the concentrations of the chemical in question differ significantly between the dump area and the control area. Furthermore, we have found no statistically significant variability among the sampling sites within areas. If we had found significant variability among the sampling sites, we would have concluded that any differences between the two areas would have been due mainly to random variability among the sites. Examples of more extensive nested designs and procedures for their analysis will be found in several of the references listed in the bibliography.

9.3 THE RANDOMIZED-BLOCK DESIGN

You should recall that the error mean square is calculated as

$$MS_{Error} = \frac{SS_{Error}}{d.f._{Error}}.$$

Now, since the F-value is computed as

$$F = \frac{MS_{Treatment}}{MS_{Error}},$$

it is apparent that a smaller error mean square gives us a better chance to detect differences among treatments, especially when those differences are small. In other words, with a reduced error mean square we will make fewer Type II errors over a period of time.

Now, if you look again at the formula for the error mean square, you can see that a reduction in the *error sum of squares* would make the error MS smaller, even if the sample size remained the same.

In certain experimental situations, the *randomized-block* design may help reduce the error SS and so increase the chance of detecting true differences among treatment means. The randomized-block design has its origins in agricultural research, where it is used to reduce the error that results from heterogeneity of environmental factors in a large experimental field. For example, in testing for differences in yield among four genetic varieties of corn, blocks can be established in the experimental area and each of the four varieties can be randomly assigned to each of four plots making up a block. This general plan is illustrated in Fig. 9.2.

In this situation, the experimenter assumes that the environmental factors *within* a block will be relatively constant, and that maximum heterogeneity of environmental factors exists *among* the blocks. In other words, the randomized-block design assumes that *minimum* variation exists *within* each block and *maximum* variation exists *among* the blocks. The total SS for error can then be reduced by computing and removing the SS that is due to variation *among* blocks. It therefore follows that proper blocking procedure involves making the experimental units *within* each block as alike as possible, leaving the greatest variation *among* the blocks.

The following example will illustrate how a randomized-block design might be used in a laboratory situation:

C	A	D	B	A	C	B	D
B	D	A	C	C	B	D	A
A	B	C	D	C	D	B	A
B	D	C	A	A	C	D	B
A	B	D	C	C	B	D	A
A	D	B	C	C	A	D	B

Fig. 9.2 Randomized block design; twelve blocks, four varieties randomly planted in each block.

Example. An investigator tested the effects of drugs A and B on the lymphocyte count in mice by comparing A, B, and a placebo, P. In designing the experiment, he assumed that mice from the same litter would be more homogeneous in their responses than would mice from different litters. He arranged the experiment in the form of randomized blocks, with three litter-mates forming each block. There was a total of seven blocks; thus there were seven replications per treatment. In each block the litter-mates were assigned randomly to the three treatments, resulting in the following arrangement, where lymphocyte counts are expressed in thousands per cubic millimeter of blood.

Treatments	Blocks (litters)							Treatment totals
	1	*2*	*3*	*4*	*5*	*6*	*7*	
P	5.4	4.0	7.0	5.8	3.5	7.6	5.5	38.8
A	6.0	4.8	6.9	6.4	5.5	9.0	6.8	45.4
B	5.1	3.9	6.5	5.6	3.9	7.0	5.4	37.4
Block totals	16.5	12.7	20.4	17.8	12.9	23.6	17.7	121.6

1. Computing the means, we obtain

$$\bar{X}_P = 5.54, \qquad \bar{X}_A = 6.48, \qquad \bar{X}_B = 5.34.$$

2. The total sum of squares is computed as usual by:

$$SS_{Total} = \sum X^2 - \frac{(\sum X)^2}{n_T}$$

$$= (5.4)^2 + (6.0)^2 + (5.1)^2 + \cdots + (6.8)^2 + (5.4)^2 - \frac{(121.6)^2}{n_T}$$

$$= 734.40 - 704.12 = 37.34.$$

3. The sum of squares for blocks is computed by using the block totals and the number of measurements (b) per block as follows:

$$SS_{Blocks} = \frac{(\sum X_1)^2}{b_1} + \frac{(\sum X_2)^2}{b_2} + \cdots + \frac{(\sum X_7)^2}{b_7} - \frac{(\sum X)^2}{n_T}$$

$$= \frac{(16.5)^2}{3} + \frac{(12.7)^2}{3} + \cdots + \frac{(17.7)^2}{3} - 704.12$$

$$= 734.40 - 704.12 = 30.28.$$

4. The sum of squares for treatments is computed by using the treatment totals and the number of replications per treatment. Thus,

$$SS_{Treatment} = \frac{(\sum X_P)^2}{n_P} + \frac{(\sum X_A)^2}{n_A} + \frac{(\sum X_B)^2}{n_B} - \frac{(\sum X)^2}{n_T}$$

$$= \frac{(38.8)^2}{7} + \frac{(45.4)^2}{7} + \frac{(37.4)^2}{7} - 704.12$$

$$= 709.33 - 704.12 = 5.21.$$

5. The sum of squares for error can now be computed by removing the sum of squares for treatment and the sum of squares for blocks from the total sum of squares. Thus,

$$SS_{Error} = SS_{Total} - (SS_{Blocks} + SS_{Treatment})$$

$$= 37.24 - (30.28 + 5.21) = 1.75.$$

6. We may now establish an analysis-of-variance table as follows:

Source	d.f.	SS	MS
Blocks	6	30.28	
Treatment	2	5.21	2.60
Error	12	1.75	0.14
Total	20	37.24	

7. Finally, we compute an F-value in the usual way by testing the MS for "treatments" over the MS for "error." Thus,

$$F_{[2,12]} = \frac{MS_{Treatments}}{MS_{Error}}$$

$$= \frac{2.60}{0.14} = 18.75,$$

which is significant well beyond the 0.01 level.

Looking back at the analysis-of-variance table, you can see that the value of the sum of squares removed from the error because of differences among the letters (blocks) was 30.28, which is a considerable reduction in the error sum of squares. Our test was therefore more sensitive to differences among the treatments than it would have been if those differences had been obscured by the meaningless but troublesome variation among the litters.

Now, go back once again to the analysis-of-variance table and note that six degrees of freedom were removed for "blocks" (there were seven blocks and the degrees of freedom for "blocks" is $b - 1$). This means that the degrees of freedom for error were reduced from 18 to 12, which is a sizable reduction. Now, suppose the investigator had been wrong in his assumption that the variability *among* litters was meaningfully larger than the variability *within* the litters. In that event, the sum of squares for blocks might have reduced the error SS by only a small amount, thus resulting in a small reduction in the error SS but a large reduction in the error degrees of freedom.

Going back once more to the following formula for the error mean square,

$$MS_{Error} = \frac{SS_{Error}}{d.f._{Error}},$$

you can see that in this design we are "playing around" with the error SS and degrees of freedom. In other words, before we decide upon a randomized-block design, we should be reasonably certain that the variability *among* the blocks is relatively large and the variability *within* each block is relatively small; otherwise we are taking a chance on uselessly decreasing the error degrees of freedom. In other words, if the error sum of squares is not significantly reduced by using the randomized-block design, the reduced error degrees of freedom will result in a larger error MS with reduced sensitivity. In this event, the *completely randomized design* would probably be more efficient.

When used correctly, however, the principle of blocking is one of the more useful designs and may be applied to a variety of biological research problems. In some cases, for example, a single organism may constitute a block, provided that two or more treatments can be applied without residual effects from one of the treatments.

9.4 FACTORIAL EXPERIMENTS

The varieties of factorial experiments that may be encountered in biological experimentation are quite extensive, and much of this topic is beyond the scope of this book. In this section, we will examine a simple two-factor design which is very useful in many experimental situations.

Essentially, a factorial experiment permits the separation and evaluation of the effects of each of two (or more) factors operating in a single experimental setting. In addition, this permits the detection of *interaction* effects between two (or more) factors. By interaction, we mean that factor A may have different effects when operating in the presence of factor B than when factor B is not present. To use a somewhat rough analogy, a young man may enjoy going to the movies, and he may enjoy going out with girls. Now, if he takes a girl to a movie, we might assume that his enjoyment of the entire evening is intensified due to the interaction between the two factors.

Or suppose that we test the effects of pH on some variable while the temperature is held constant. Obviously, it will be more enlightening if we can determine whether the effects of pH are different at different temperatures.

Example. An experimenter wished to determine the effect of a drug versus a placebo relative to blood pressure in human subjects and was also interested in the effect that would be produced by a possible interaction with the factor *sex*. The experimenter therefore set up a two-factor factorial experiment with five replications per cell, as in Table 9.3.

1. Our first step is to find the totals shown in the table. These include (a) cell totals, (b) total for males, (c) total for females, (d) totals for drug and placebo, and (e) the usual $\sum X^2$ and $\sum X$ that are found by utilizing all measurements found in the data. The symbols shown have appropriate subscripts which identify the particular factor level or combinations of levels represented.

2. We begin the analysis by computing the total sum of squares by:

$$SS_{Total} = \sum X^2 - \frac{(\sum X)^2}{n_T} = 417{,}322 - \frac{2872^2}{20}$$

$$= 417{,}322.00 - 412{,}419.20 = 4902.80.$$

TABLE 9.3

Sex factor	Drug factor		
	Placebo	Drug	
Male	153 140 133 $\sum X_{mp} = 712$ 123 163	132 115 142 $\sum X_{md} = 668$ 125 154	$\sum X_m = 1380$
Female	164 150 134 $\sum X_{fp} = 766$ 144 174	142 155 167 $\sum X_{fd} = 726$ 133 129	$\sum X_f = 1492$

$$\sum X_p = 1478 \qquad \sum X_d = 1394$$
$$\sum X^2 = 417{,}322$$
$$\sum X = 2872$$

3. Next, we compute the sum of squares for the drug factor by utilizing the total for the placebo and the total for the drug. Thus,

$$SS_{Drug} = \frac{(\sum X_p)^2}{n_p} + \frac{(\sum X_d)^2}{n_d} - \frac{(\sum X)^2}{n_T}$$

$$= \frac{1478^2}{10} + \frac{1394^2}{10} - 412{,}419.20$$

$$= 412{,}772.00 - 412{,}419.20 = 352.80.$$

4. We now compute the sum of squares for the sex factor, using the male and female totals. Thus,

$$SS_{Sex} = \frac{(\sum X_m)^2}{n_m} + \frac{(\sum X_f)^2}{n_f} - \frac{(\sum X)^2}{n_T}$$

$$= \frac{1380^2}{10} + \frac{1492^2}{10} - 412{,}419.20$$

$$= 413{,}046.40 - 412{,}419.20 = 627.20.$$

5. Our next step is the computation of the sum of squares for the effect of interaction. To do this we use the totals for the various cells, since each

cell total represents a combination of a specific level of one factor with a specific level of the other. Thus

$$SS_{Interaction} = \frac{(\sum X_{mp})^2}{n_{mp}} + \frac{(\sum X_{md})^2}{n_{md}} + \frac{(\sum X_{fp})^2}{n_{fp}} + \frac{(\sum X_{fd})^2}{n_{fd}} - \frac{(\sum X)^2}{n_T}$$

$$= \frac{712^2}{5} + \frac{668^2}{5} + \frac{766^2}{5} + \frac{726^2}{5} - 412,419.20$$

$$= 413,400 - 412,419.20 = 980.80.$$

Since the sum of squares computed for interaction is "contaminated" by the separate effects due to the drug and sex factors, we must now remove these effects by removing the sums of squares previously computed for drug and sex. Thus

$$SS_{Interaction} = 980.80 - (352.80 + 627.20)$$

$$= 980.80 - 980.00 = 0.80.$$

6.   Finally, the sum of squares for error may be computed by direct subtraction, such as:

$$SS_{Error} = SS_{Total} - (SS_{Drug} + SS_{Sex} + SS_{Interaction})$$

$$= 4902.80 - (352.80 + 627.20 + 0.80)$$

$$= 4902.80 - 980.80 = 3922.00.$$

7.   We may now set up an analysis-of-variance table as in Table 9.4.

TABLE 9.4

Source	d.f.	SS	MS
Drug	1	352.80	352.80
Sex	1	627.20	627.20
Interaction	1	0.80	0.80
Error	16	3922.00	245.12
Total	19	4902.80	

Note that degrees of freedom for the main effects are found by the usual method. There is a new factor to consider—that of interaction. For interaction, the degrees of freedom are computed by multiplying together the degrees of freedom for the main effects. Thus, in this case, $1 \times 1 = 1$.

We may now perform the appropriate F-tests as follows:

$$F_{1,16(\text{Treatment})} = \frac{352.80}{245.12} = 1.43,$$

$$F_{1,16(\text{Sex})} = \frac{627.20}{245.12} = 2.55,$$

$$F_{1,16(\text{Interaction})} = \frac{0.80}{245.12} = 0.0032.$$

8. Turning to Table VII, it may be seen that the F-value for the main effects, sex and drug, do not reach significance at the 0.05 level. The F-test for interaction is less than one and may therefore be automatically regarded as nonsignificant. We may therefore conclude that the results of this experiment indicate no statistical significance for the drug effect, the effect of sex, or the interaction effect.

In this chapter we have presented three of the basic designs made possible by analysis of variance. There are a number of interesting and extremely useful extensions and variations on the basic experimental procedures examined here, and the reader should look to the bibliography for some excellent references on the subject.

PROBLEMS

9.1 An experiment was performed to test the effects of a certain drug on blood coagulation time (in minutes). Eight subjects were randomly divided into two groups of four each. One group was given the drug and the other group was used as a control. The blood-coagulation time was determined on each of four replicate samples obtained from each of the eight subjects. Analyze the following data for (a) statistically significant difference between the drug and the control, and (b) statistically significant variability among the subjects. (See Section 9.2.)

	Drug					Control			
Subjects					Subjects				
A	1.6,	1.8,	1.9,	2.0	E	1.8,	2.0,	2.1,	1.7
B	1.9,	2.1,	1.6,	1.8	F	1.7,	1.8,	2.1,	1.9
C	1.3,	1.6,	1.4,	1.4	G	2.2,	2.3,	2.1,	2.3
D	1.5,	1.2,	1.6,	1.5	H	2.1,	1.9,	2.2,	1.8

9.2 An experiment was performed to see whether two different brands of counting chambers produced significantly different platelet counts. Six experienced technicians were selected; three of the technicians were randomly assigned to use chamber A and the other three used chamber B. Each of the six technicians was given a blood sample from the same source; and they were instructed to make three replicate platelet counts from the sample. The following data are the platelet counts in thousands per cubic millimeter of blood. Analyze these data for statistical significance between the counting chambers and for variability among the technicians. (See Section 9.2.)

Chamber A

Count	Technicians		
	T_1	T_2	T_3
First	248	272	260
Second	252	247	255
Third	243	259	253

Chamber B

Count	Technicians		
	T_4	T_5	T_6
First	242	262	250
Second	251	257	242
Third	246	259	243

9.3 Three clinical methods of determination of hemoglobin content were tested for significant difference in results. Six subjects were used, each subject constituting a single block. Analyze the date in Table 9.5, where the figures represent grams/100 ml (see Section 9.3).

TABLE 9.5

Method	Blocks					
	A	B	C	D	E	F
1	14	12	16	15	10	11
2	18	16	17	19	12	13
3	15	14	12	14	12	9

9.4 Two diets were tested, using hamsters, for significant difference in terms of final weight measured after a specified period of time. The subjects were blocked, two to a block, the blocking done on the basis of predict weight. Test the data in Table 9.6 for significant difference between diets. The weights are expressed in grams (see Section 9.3).

TABLE 9.6

Diet	Blocks									
	1	*2*	*3*	*4*	*5*	*6*	*7*	*8*	*9*	*10*
A	105	101	103	108	106	109	105	106	104	103
B	110	108	106	112	110	112	110	106	108	108

9.5 Five samples of plankton were taken from each of two locations on a lake during the month of May. This process, using the same locations, was repeated in early August. The data, expressed in thousands of plankton per liter, are given below. Test these data for significant difference between locations, between collection times, and also for interaction between location and time (see Section 9.4).

Location I	August (A)	97,	103,	109,	98,	102
	May (B)	108,	113,	119,	109,	112
Location II	August (A)	106,	110,	116,	105,	111
	May (B)	111,	116,	120,	111,	113

9.6 An experiment was performed to test the photoreactivation phenomenon found in bacteria that have been exposed to ultraviolet. A plate of bacteria was exposed to ultraviolet for five minutes. A sample was taken from the exposed plate, plated, exposed to visible light for five minutes, then incubated. A second sample was exposed to visible light for a ten-minute period before incubation. A third sample was incubated without prior exposure to visible light. After a specific incubation time, the colonies were counted. The following data represent colony counts from five replications per treatment. Analyze for significant difference in number of colonies (see Sections 8.6 and 8.7).

No exposure to visible light—45, 40, 10, 23, 32
Five minutes exposure to light—72, 81, 53, 55, 48
Ten minutes exposure to light—90, 100, 75, 70, 64

9.7 Three strains of *Drosophila pseudoobscura* were bred for resistance to an insecticide. Three levels of concentration of the insecticide were tested with the three inbred strains. The data, expressed in percent of mortality over a period of time, are based on five replications per treatment combination. Analyze these data for significant differences in mortality rate among strains, and among levels of insecticide, and for interaction between the strain and the insecticide. (*Note:* Because the data fall between 30% and 80%, it will not be necessary to use the arcsin transformation.) (See Section 9.4.)

Insecticide level 1
Strain A	60,	55,	52,	38,	31
Strain B	58,	53,	50,	35,	30
Strain C	37,	43,	·57,	60,	66

Insecticide level 2
Strain A	44,	37,	54,	57,	65
Strain B	63,	59,	54,	38,	38
Strain C	59,	51,	53,	62,	71

Insectidice level 3
Strain A	46,	51,	63,	66,	74
Strain B	63,	44,	46,	66,	71
Strain C	51,	80,	68,	71,	55

CHAPTER TEN

Correlation and Regression

10.1 INTRODUCTION

In biological situations we often find two or more variables that are apparently associated or interdependent. For example, we may have a situation in which an increase in variable X is accompanied by a corresponding increase in variable Y, or a decrease in X appears related to a decrease in Y. When it can be demonstrated that the variation in one variable is somehow associated with the variation in another variable, then the two variables may be said to be *correlated*. A correlation may be *positive*, as when X and Y both increase, or *negative*, as when one variable increases while the other decreases. On the other hand, if the variation in X and the variation in Y do not correspond at all, then there is *no* association and therefore *no correlation* between the two variables.

A common error is the assumption that because X and Y show a high correlation, it follows that one must be the cause of the other. For example, consider the high correlation between the consumption of cigarettes and lung cancer. Do we automatically conclude that smoking cigarettes *causes* lung cancer? If so, then we should also consider that a high correlation exists between the incidence of lung cancer and the number of telephone poles. Should we therefore conclude that telephone poles cause lung cancer? Or what about the correlation between lung cancer and the number of automobiles? Finally, does the high correlation between the number of drownings on a given day and the number of ice cream cones sold on that day indicate that those who eat ice cream cones are more likely to drown?

As usual, our interpretation of statistics must be leavened with common sense and practical biological considerations. As far as can be determined, no one is exposed to carcinogens by walking in the vicinity of a telephone pole. Historically, however, the growth in the number of telephone poles has kept pace with the steady increase in the consumption of cigarettes. Cigarettes *do* contain certain carcinogens, as do automobile exhaust fumes.

In brief, a high correlation between two variables does not necessarily imply causation. A significant correlation may suggest such a possibility to an investigator, but supportive evidence from other sources must be found before a conclusion that involves causation is warranted.

One of the objectives of this chapter is to distinguish between correlation and regression analysis. There is a distinction between these two procedures and one should not be substituted for another in a given experimental situation. Quite often, this distinction is not made clear; and investigators are sometimes confused as to which procedure to use.

Regression analysis involves the measurement of the "degree of dependency" of a dependent variable (Y) on an independent variable (X). Thinking back to Chapter 1, you will recall that the independent variable is manipulated by the experimenter. In other words, the experimenter decides which values the independent variable shall take, and the values of the dependent variable are determined by the relationship, if any, that exists between the dependent and independent variables. Therefore, regression analysis should be used in experimental situations in which the investigator is controlling the independent variable.

For example, if an investigator measures the degree of dependency of heart rate in alligators by subjecting alligators to specific temperatures such as $10°$, $15°$, $20°$, $25°$, etc., while observing the heart rate at those *specified* temperatures, the situation calls for regression analysis. In this case, temperature is *not* a random variable because definite temperature values are established by the investigator. Heart rate, on the other hand, *is* a random variable, and it is not under the investigator's control. As we will see in later sections of this chapter, regression analysis can be used to *predict* the value of Y that will result from the application of a specific value of X.

Now, suppose an investigator wants to determine the degree of association that exists between the biomass of phytoplankton and the amount of chlorophyll a. The investigator therefore draws replicate samples of water from a sampling site on a lake and measures the chlorophyll a and biomass in each replicate sample. In this situation, the investigator has no control over either variable, since the values of both chlorophyll a and biomass found in each sample will be "what nature provides." It therefore follows that chlorophyll a and biomass are both random variables, and correlation is therefore the appropriate statistical procedure.

We should also point out that, since both chlorophyll a and biomass are, in a sense, dependent variables, it is not strictly correct to designate them by the symbols X and Y, because X should be used to denote an independent variable. However, because these symbols are so deeply chiseled in the stone of tradition, we will continue to use them in our discussion of correlation. It should be understood, however, that where correlation analysis is concerned, neither X nor Y represents an independent variable; in other words, both X and Y are random variables. On the other hand, when we discuss regression analysis, X will denote the independent variable and Y the dependent variable.

10.2 THE CORRELATION COEFFICIENT

Correlation is expressed by a coefficient (r) which may range from -1 to $+1$. A coefficient of 1, positive or negative in sign, indicates a perfect correlation between two variables. On the other hand, a coefficient of zero suggests a complete lack of correlation. Varying degrees of correlation are then represented by coefficients ranging from zero to 1 in either direction.

We can plot a series of points representing the magnitudes of each pair of X and Y variables and obtain various patterns, some of which are shown by Fig. 10.1. Diagram (a) shows a series of points that are not on a straight line. This represents a correlation that is less than perfect, but shows a general trend in a positive direction. If we correlated height and weight in a sample of adult human subjects, we would probably obtain a pattern similar to that shown by (a). A similar pattern of imperfect correlation is shown by (b), but in this case the slope is negative, indicating a negative correlation. A pattern similar to (b) might be obtained by plotting the incidence of dental cavities in children against the level of fluoridation. Diagram (c) shows a perfect negative (-1) correlation, such as might be obtained from plotting acceleration against mass in the application of Newton's second law, $F = ma$. Diagram (d) shows the random scattering of points that might be expected from plotting Y against X when the two variables are not at all associated.

What is a "good" correlation? This obviously depends on what the investigator is doing, or what he hopes to find. If one hopes to show that no association exists between two variables, then a coefficient of zero would be a cause for celebration. If, on the other hand, the investigator hopes to show a strong association, then an r approaching 1 would be considered optimum. As usual, living systems do not usually cooperate by producing nice round figures; variability is the rule rather than the exception. In most cases, therefore, the biologist will probably need to be content with coefficients that are less than perfect.

Squaring the correlation coefficient (r) yields the *coefficient of determination* (r^2). This value may be used as an estimate of the intensity of association between the two variables that appear to be correlated. Specifically, the coefficient of determination estimates the percentage of

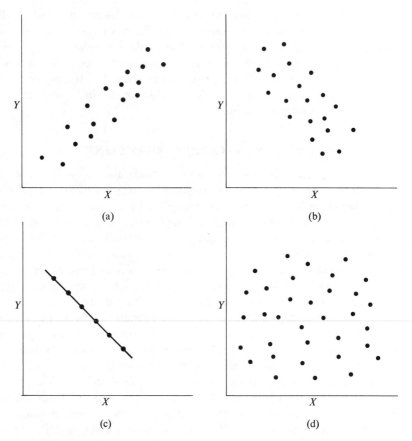

Fig. 10.1 Typical patterns obtained from plotting Y against X.

the variation in X that is associated with (or "explained by") the variation in Y—or vice versa. For example, if the sample correlation between two variables such as chlorophyll a and biomass is 0.50, squaring this coefficient yields a coefficient of determination of 0.25. This suggests that 25 percent of the variation in one of the two variables is associated with or "explained by" the variation in the other. We cannot say which "explains" which because they are both considered to be dependent variables.

Of course, when we say that 25 percent of the variation in one of two variables is "explained" by the other, we are taking the optimist's point of view that a glass is one-quarter full. The pessimist, on the other hand, would be quick to point out that the glass is three-quarters empty! In a similar way, our case does not sound nearly as strong when we say that 75 percent of the variation in one of the two variables is *not* explained by the variation in the other.

If we look at the arithmetic of the situation, we can see that the value of r^2 increases rapidly as r increases. For example, when $r = 0.10$, $r^2 = 0.01$; when $r = 0.20$, $r^2 = 0.04$, when $r = 0.60$, $r^2 = 0.36$; when $r = 0.90$, $r^2 = 0.81$, and so on. It therefore follows that we have to obtain a reasonably large value of r in order to claim a high degree of association between two variables.

10.3 COMPUTING THE CORRELATION COEFFICIENT (r)

Consider the following simple example of a bivariate distribution:

Pair	X	Y
1	1	6
2	2	7
3	3	8
4	4	9
5	5	10

It is obvious from a simple inspection of these data that a perfect positive correlation exists between X and Y. Now, if we compute $\overline{X}$ and $\overline{Y}$, and the standard deviations of the X- and Y-distributions, we may use these quantities to compute the Z-value of each variate in the usual manner,

$$Z_X = \frac{X - \overline{X}}{S_X} \quad \text{and} \quad Z_Y = \frac{Y - \overline{Y}}{S_Y}.$$

This would yield distributions of signed Z-values as shown in Table 10.1.

TABLE 10.1

Pair	X	Y	$Z_X Z_Y$
1	$-Z_X$	$\times\ -Z_Y$	$= +Z_X Z_Y$
2	$-Z_X$	$\times\ -Z_Y$	$= +Z_X Z_Y$
3	0	$\times\ 0$	$= 0$
4	$+Z_X$	$\times\ +Z_Y$	$= +Z_X Z_Y$
5	$+Z_X$	$\times\ +Z_Y$	$= +Z_X Z_Y$

$$r = \frac{\sum Z_X Z_Y}{n} = +1$$

Note that all cross products of Z-values are plus, and if the Y-distribution exactly "fits" the X-distribution, the Z_X- and Z_Y-values will be the same for both members of each pair. Summing the cross products of the Z-values and dividing by n, the number of pairs, yields the correlation coefficient 1. We can see that r is basically the mean of all cross products of Z-values found in the bivariate distribution.

Now consider the bivariate distribution shown in Table 10.2.

TABLE 10.2

Pair	X	Y
1	1	10
2	2	9
3	3	8
4	4	7
5	5	6

This time we can also see a perfect relationship, but in the opposite direction. In other words, the correlation is negative. Again computing Z-values, we obtain the distributions shown in Table 10.3.

TABLE 10.3.

Pair	X	Y	$Z_X Z_Z$
1	$-Z_X$	$\times +Z_Y =$	$-Z_X Z_Y$
2	$-Z_X$	$\times +Z_Y =$	$-Z_X Z_Y$
3	0	$\times\ 0\ =$	0
4	$+Z_X$	$\times -Z_Y =$	$-Z_X Z_Y$
5	$+Z_X$	$\times -Z_Y =$	$-Z_X Z_Y$

$$r = \frac{\sum Z_X Z_Y}{n} = -1$$

Calculating the cross products of Z-values of each pair, we obtain a $Z_X Z_Y$ sum which has a negative sign. Dividing by n, we then obtain -1 as the coefficient. Again, the correlation is perfect, but the Y-distribution is turned end to end before it is superimposed on the X-distribution.

It follows logically that a bivariate distribution in which X- and Y-values do not correlate perfectly will yield $Z_X Z_Y$-values that are signed in a more irregular, haphazard manner. This will yield a mean $Z_X Z_Y$-value ranging downward from $+1$ (or upward from -1), to zero in cases where there is no association at all.

Now, given any significant amount of data, this method of computation would be prohibitively tedious, so we need to develop an easier approach to the calculation of r.

Recalling the formula for computing a Z-value, it may be seen that the cross product, $Z_X Z_Y$, is the same as

$$\frac{X - \bar{X}}{S_X} \times \frac{Y - \bar{Y}}{S_Y},$$

and therefore $\sum Z_X Z_Y$ may be presented by

$$\frac{\sum (X - \bar{X})(Y - \bar{Y})}{n S_X S_Y},$$

and some relatively simple algebraic manipulation yields

$$r = \frac{\sum XY - (\sum X \sum Y / n)}{\sqrt{(\sum X^2 - (\sum X)^2 / n)(\sum Y^2 - (\sum Y)^2 / n)}},$$

which is a useful computing formula for finding r, the correlation coefficient. To the biologist used to counting the number of legs on a grasshopper, this formula may appear to be rather formidable; but with a reasonably good desk calculator the values $\sum XY$, $\sum X^2$, $\sum Y^2$, $\sum X$, and $\sum Y$ can be easily obtained with one "run" through a set of bivariate data.

10.4. TESTING HYPOTHESES ABOUT ρ

Suppose that we draw a sample from a bivariate population and compute r. Suppose further that our sample r turns out to be 0.45. How do we know whether our sample r represents a true correlation between the X and Y variates of the *population*, or is only a chance deviation from zero? Actually, if we compute a sample correlation between two variables drawn from almost any source, it would be unusual to obtain a coefficient of *exactly* zero, even if no correlation whatsoever existed between all the X and Y variates in the population itself. This would be similar to tossing 1000 coins in the air and obtaining exactly 500 heads! It's possible, but not likely!

We are therefore back to the familiar problem of determining whether a sample statistic is statistically significant, or if it occurred by chance alone. Figure 10.2 shows a theoretical sampling distribution based on r's associated with an infinitude of samples drawn from a population in which the *parameter* value of the correlation coefficient is zero. In keeping with our established procedure, we let the Greek letter ρ (rho) represent the population coefficient. From the sampling distribution, we note that the sample coefficients (r) could theoretically range anywhere from -1 to $+1$, but most of the samples obtained would yield coefficients clustering around the mean of zero. We are therefore saying that, while the probability is admittedly very small, it is nevertheless *possible* to draw a sample from a bivariate population where no correlation whatsoever exists between the variables

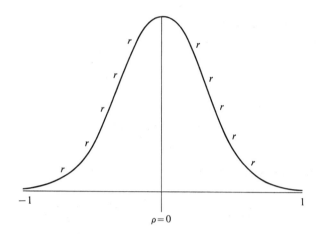

Fig. 10.2 Sampling distribution of r, based on population in which ρ is zero.

and, by chance alone, find a perfect correlation between the sample variates. It may also be seen that a sample yielding an r of zero is highly improbable, and we are therefore not surprised if a sample drawn from such a population shows a correlation other than zero! The following problem demonstrates a method of determining whether a sample coefficient is statistically significant or is simply a chance deviation from zero.

Example.

1. A sample of $n = 42$ pairs of measurements is randomly drawn from a population. A correlation coefficient is computed and yields $r = 0.60$. Does this indicate that a real correlation exists between the two variables in the population, or is the sample r-value only a chance sampling deviation from zero?

2. Since we are interested in whether or not the *population* coefficient differs significantly from zero, we write the null hypothesis as

$$H_0: \rho = 0.$$

3. We will set a level of significance of 0.05. Also, we will assume that we are not concerned about the *direction* of difference, so we will use a two-tailed test. Figure 10.4 shows a sampling distribution of r's based on a population where $\rho = 0$.

4. As usual, we must place our statistic ($r = 0.60$) on the sampling distribution. In this case we will use a practical short-cut formula that will

yield a *t*-value as follows:

$$t = \frac{r - 0}{\sqrt{(1 - r^2)/(n - 2)}} = \frac{0.60}{\sqrt{(1 - (0.60)^2)/(42 - 2)}} = 4.76.$$

5. We now enter Table III at 40 degrees of freedom and find the critical value of *t* at the 0.05 level for a two-tailed test. We find that our *t*-value of 4.76 exceeds the critical value at the 0.05 level and also exceeds the critical value for the 0.001 level of significance. Note that we entered Table III with $(n - 2)$ degrees of freedom. This is indicated in the above formula and we can also think in terms of using two parameter estimates (the mean of X and the mean of Y) in our computations of *r*. At this point you should recall the general rule that we lose a degree of freedom for each parameter estimate used in our computations.

6. Since our *t*-value exceeds the critical value at the 0.001 level, we reject the null hypothesis ($H_0: \rho = 0$) beyond the 0.001 level. We therefore have statistical evidence that a true correlation between the X and Y variables exists in the population. In other words, we can say that our sample correlation of 0.60 is *significant*. At this point, however, we must be extremely careful *not* to confuse *significant* with *meaningful, useful*, etc. Another look at the above formula will show that a relatively small value of *r* can be significant in the *statistical* sense if the *n*-number is large enough. Whether or not such a correlation would be meaningful or useful is quite another matter.

10.5 ESTIMATING ρ

Suppose we find that a particular sample *r* of 0.45 is significant, and therefore indicates a real correlation between the population X- and Y-variables. At this point it would be an error to assume that the parameter ρ is also 0.45. This would be similar to concluding that μ is, say, 50, because we have drawn a sample yielding an $\overline{X}$ of 50.

As before, it is necessary to estimate the population parameter with confidence limits, using the sample statistics as a basis for the estimation. Before we proceed with this operation, however, it will be necessary to discuss the nature of the distributions that may be obtained relative to ρ.

Looking back at Fig. 10.2, it may be seen that when ρ is zero, the sampling distribution of ρ is symmetrical, since the sample *r*'s distribute normally from -1 to $+1$, with the majority of cases clustering around $\rho = 0$. Suppose, on the other hand, that ρ equals 0.90. From a population such as this, it is apparent that we would obtain a preponderance of samples having high values of *r*, with fewer samples having low or negative values. This would therefore produce a skewed distribution, as shown in

Fig. 10.3(a). Another distribution, based on $\rho = 0.10$, which, as might be expected, is skewed in the other direction, is shown in Fig. 10.3(b).

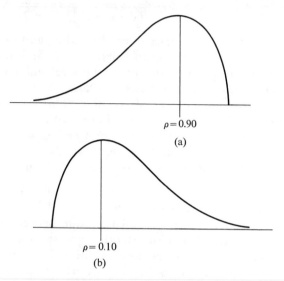

$\rho = 0.90$

(a)

$\rho = 0.10$

(b)

Fig. 10.3 Sampling distributions of r drawn from populations where $\rho = 0.90$ and $\rho = 0.10$.

Since we cannot apply normal-curve principles to a badly skewed distribution, it is necessary to apply a transformation when working with correlations that are significantly greater or less than zero. The only time this transformation is *not* necessary is when dealing with distributions based on $\rho = 0$, as in Section 10.4. Fortunately, the transformations involving r that normalize distributions of r that would otherwise be skewed have been worked out for us by R. A. Fisher, who obtained the so-called r-to-Z values shown in Table XI. Looking at Table XI, it is evident that the farther r is from zero, the more necessary it is to use the r-to-Z transformation.

Suppose that we demonstrate both the estimation of ρ and the use of the r-to-Z transformation with the following example.

Example. A sample of 65 pairs of variates was drawn from a population. Computation of the correlation coefficient yielded $r = 0.45$. Estimate the population coefficient ρ from the sample data, using 95 percent confidence limits.

1. First, we must convert r to Z. Consulting Table XI, we find that a correlation of 0.45 is equivalent to a Z-value of 0.485.

2.   Next, we need to compute the standard error of Z. This is given by

$$S_Z = \frac{1}{\sqrt{n-3}} = \frac{1}{\sqrt{62}} = \frac{1}{7.89} = 0.127.$$

3.   Now we can proceed as usual with the estimation by

$$Z \pm S_Z(1.96) = 0.485 \pm 0.127(1.96),$$

$$0.485 \pm 0.249 = 0.236\text{–}0.734.$$

4.   Finally, converting back from Z to r, using Table XI, we find that the confidence interval for ρ is

$$0.23 \leq \rho \leq 0.62.$$

5.   We may therefore make the statement that, based on our sample data, we estimate that ρ falls between 0.23 and 0.62, and we are confident of being correct in this estimate 95 percent of the time.

10.6 AN EXAMPLE OF CORRELATION ANALYSIS

In this section we will follow through a complete correlation analysis, using the procedures described earlier in this chapter.

Suppose we are interested in finding out whether there is an association between body weight and blood-cholesterol concentration. We begin by randomly selecting 15 subjects from a population of adult males between the ages of 50 and 55, with a height between 69 and 70 inches. We then measure the body weight and the concentration of cholesterol for each subject, obtaining the data shown in Table 10.4. Although weight and cholesterol content are both obviously random variables, for purposes of convenience we will assign the traditional X and Y symbols to the variables. It should be stressed, however, that in this case weight is *not* an independent variable.

Computation of r

Computer programs for the calculation of the correlation coefficient and the various tests associated with correlation are readily available. The computer may therefore be the choice of the investigator who is faced with large masses of data. Otherwise, there are now available a variety of desk calculators, including some rather sophisticated statistical calculators. For example, the entire analysis presented in this section was completed in about four minutes with a relatively inexpensive statistical calculator.

When using a calculator, we can make our work somewhat easier by using a coding procedure to reduce the size of the numbers making up the

TABLE 10.4

Subject	Weight (lbs)	Cholesterol (mg/100 ml)	Coded data	
	X	Y	X	Y
1	146	181	46	81
2	205	228	105	128'
3	157	182	57	82
4	165	249	65	149
5	184	259	84	159
6	153	201	53	101
7	220	339	120	239
8	181	224	81	124
9	151	112	51	12
10	188	241	88	141
11	181	225	81	125
12	163	223	63	123
13	198	257	98	157
14	193	337	93	237
15	157	197	57	97

data. For example, in the present situation we will subtract 100 from each of the weight and cholesterol measurements and work with the coded data shown in Table 10.4. You should recall at this time that the addition (or subtraction) of a constant to (or from) the members of a distribution produces a distribution with the *same variance* as the original distribution. In calculating r, we are, in a sense, trying to match the variances of the distributions of weight and cholesterol content; therefore, it makes no difference whether we use the original figures or the coded version.

1. Our first step is to use a calculator to obtain the following bits and pieces of data from the coded "bivariate" distribution.

$$n = 15 \qquad \sum X = 1142 \qquad \sum Y = 1955$$
$$\sum XY = 162{,}808 \qquad \sum X^2 = 93{,}818 \qquad \sum Y^2 = 300{,}855$$

2. We can now substitute the above values in the machine formula for r as follows:

$$r = \frac{\sum XY - (\sum X \sum Y/n)}{\sqrt{\sum X^2 - [(\sum X)^2/n])(\sum Y^2 - [(\sum X)^2/n])}}$$

$$= \frac{162{,}808 - [(1142)(1955)/15]}{\sqrt{(93{,}818 - [(1142)^2/15])(300{,}855 - [(1955)^2/15])}}$$

$$= 0.78.$$

Testing r for Statistical Significance

Having obtained a sample correlation of 0.78, we now use the procedure described in Section 10.4 to test whether or not this sample coefficient reflects a real correlation in the population. In other words, is our sample coefficient of 0.78 only a chance deviation from a population ρ that is actually zero? We therefore test the null hypothesis, $H_0: \rho = 0$ by using the following formula:

$$t = \frac{r - 0}{\sqrt{(1 - r^2)/(n - 2)}} = \frac{0.78}{\sqrt{(1 - 0.61)/13}} = 4.51.$$

Entering Table III at $(n - 2)$, or 13 degrees of freedom, we find that r is statistically significant beyond the 0.001 level, using a two-tailed test. Again, we must emphasize *statistically significant*; we are not *necessarily* using "significant" as a synonym for "meaningful" or "useful," even though this is an error often found in the journals.

Estimating ρ

Having found that the sample correlation coefficient is statistically significant beyond the 0.001 level, we can legitimately suspect that a real correlation of some kind exists between body weight and blood cholesterol in the population of interest. On the other hand, at this point in our development as statisticians we would be unlikely to suggest that the true population coefficient (ρ) is 0.78 just because the sample coefficient is 0.78. As usual, therefore, we must *estimate* the population value, this time by following the procedure discussed in Section 10.5. Thus, to estimate ρ with 95 percent confidence, we use:

$$Z \pm \frac{1}{\sqrt{n - 3}}(t_{0.05}).$$

After converting 0.78 to the Z-transformation obtained from Table XI and finding the t-value for the two-tailed test at the 0.05 level with 12 degrees of freedom, we have

$$1.045 \pm \frac{1}{\sqrt{12}}(2.179),$$

or

$$0.416 \quad \text{to} \quad 1.674.$$

Converting these Z-values back to the approximate corresponding r-values (Table XI), we find that 95 percent confidence limits for ρ are:

$$0.39 \leq \rho \leq 0.93.$$

Conclusions

The coefficient of determination (r^2) associated with our sample correlation $r = 0.78$ is $r^2 = 0.61$. It is therefore tempting to claim that 61 percent of the variation in cholesterol content is "explained" or "accounted for" by the variation in body weights, or vice versa (since they are both dependent variables). This would be misleading, however, because we do *not* know that the true population coefficient is 0.78. Based on our data, the only statement we can make is that we are 95 percent confident that the true population coefficient (ρ) could be anything from 0.39 to 0.93. While it might be nice if we could claim 0.93 as the value of ρ, we must at the same time admit that ρ could be as low as 0.39. If this is true, then squaring 0.39 tells us that only 15 percent of the variation in either of the two variables could be "explained" by the other. Or, to take the pessimist's point of view, 85 percent of the variation in one is *not* "explained" by the other if we consider the possibility of the *lower* unit.

As you can see, it is wise to beware of research articles that report a sample correlation together with the cryptic statement that it "proved significant" at one level or another. Before rushing to congratulate the author, we should insist on knowing the sample size and the results of an estimation of ρ that *should* have been computed and reported.

Another lesson that we may take from this example is the fact that it is difficult to compute a meaningful or useful correlation coefficient if we do not have an adequate n-number. In our example we used 15 pairs of measurements; this small sample is responsible for the wide range of the confidence limits when we estimated ρ. Looking back at that estimation, it may be seen that if n is increased, the values of both $1/\sqrt{n-3}$ and $t_{0.05}$ will decrease. For example, if we had used a sample of 200 subjects instead of 15, and if we assume the same sample coefficient (0.78), the 95 percent confidence limits for ρ would have been

$$0.69 \le \rho \le 0.85,$$

which is a much more precise and therefore much more useful estimation. It is quite obvious, therefore, that the experimenter who hopes to make claims based on a correlation procedure must consider the need for adequate sample size as an important part of the experimental design.

The correlation coefficient that we have discussed in this chapter is a parametric statistic, and it is called the Pearson r. It assumes that both variables in the bivariate distribution are normally distributed in the population. In certain situations where it is obvious that the assumption of normality cannot be met, the Spearman rank correlation procedure discussed in Chapter 12 may be used as an alternative. Furthermore, we have considered only those procedures that are involved with a correlation between *two* variables. Multiple and partial correlations are beyond the scope of this book, but references to these procedures can be found in the bibliography.

10.7 REGRESSION ANALYSIS

Correlation analysis provides the biologist with a measure of the intensity of association between two *random* variables, neither of which is under the control of the experimenter. On the other hand, if the investigator controls, or manipulates, one of two variables, the variable that is manipulated is a true independent variable (X). In this situation, *regression*, not correlation, analysis should be performed. Regression analysis will enable the experimenter to know how much change in Y can be expected as a result of a unit change in X. Therefore, regression analysis allows us to *predict* the value of the dependent variable (Y) that is associated with a specific value of the independent variable (X).

If a perfect linear relationship exists between Y and X, the data will usually provide an easy and obvious basis for prediction. Unfortunately, examples of perfect relationships between X and Y (such as the one shown in Fig. 10.4(a)) are exceptions rather than the rule in the world of living systems. Figure 10.4(a) could represent, say, the number of annual rings in a tree versus years of growth. Most of the time, the biologist must deal with more variable data such as those shown in Fig. 10.4(b). In such situations we must fit a *regression line*, or "line of best fit," to the data.

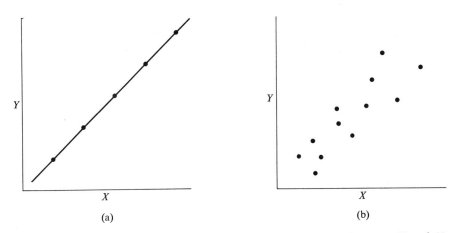

(a)   (b)

Fig. 10.4 Scatter diagrams showing two types of relationships between X and Y.

Figure 10.5(a) shows a scatter diagram of Y on X consisting of five points. A horizontal line, representing the mean of Y-values ($\overline{Y}$), is drawn

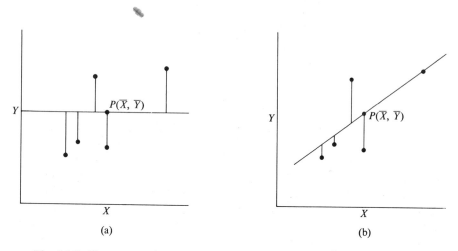

Fig. 10.5 Placement of regression line so that $\sum(Y_p - Y)^2$ is at a minimum.

through point P, which represents the point plotted from $\overline{X}$ and $\overline{Y}$, the means of the X and Y distributions. At this time it might be well to point out that the regression line *will always pass through* the point representing $(\overline{X}, \overline{Y})$.

Now, suppose that we were able to move the line in Fig. 10.5(a) around the axis formed by point P, until it reaches the position depicted in Fig. 10.5(b). Now it may be seen that the distances representing the deviations of Y from the line are decreased. This new line, called the *regression line*, represents the Y-values that would then be *predicted* from X. It is apparent that these predicted values, which we will call Y_p, would still deviate in many cases from the *actual* Y-values obtained from the data. However, the regression line has been placed so that *the sum of the squared deviations of predicted Y's from actual Y's is at a minimum.* Symbolically, $\sum(Y_p - Y)^2$ is at a minimum. This general method of plotting a "line of best fit" is called the *method of least squares.*

Figure 10.6 shows a straight line plotted on a pair of coordinates. You will recall, from elementary algebra, that we have to know two values in order to plot a straight line on a set of coordinates. One of these values is the slope (b) and the other value (a) is the Y-intercept. Knowing these, we can utilize the familiar linear equation,

$$Y_p = a + bX$$

in order to plot the line. In working with regression, however, we will need to use the data from our bivariate distribution in order to obtain a and b. The mathematician has derived the value of b for us by using the methods

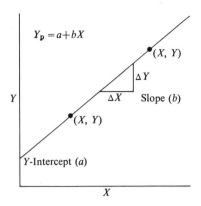

Fig. 10.6 Straight line representing basic linear equation.

of differential calculus. Thus, the slope of the line may be computed by

$$b_{YX} = \frac{\sum (X - \bar{X})(Y - \bar{Y})}{\sum (X - \bar{X})^2}.$$

A more convenient computation form is

$$b_{YX} = \frac{\sum XY - (\sum X \sum Y/n)}{\sum X^2 - [(\sum X)^2/n]}.$$

The Y-intercept, a, depends on the values $\bar{X}$ and $\bar{Y}$, and is given by

$$a = \bar{Y} - b\bar{X}.$$

This expression may be substituted for a in the basic linear equation, as follows:

$$Y_p = a + bX = \bar{Y} - b\bar{X} + bX$$
$$= \bar{Y} + bX - b\bar{X} = \bar{Y} + b(X - \bar{X}).$$

We now have the regression equation in its most useful form as

$$Y_p = \bar{Y} + b(X - \bar{X}),$$

since we can easily compute $\bar{Y}$ and $\bar{X}$, and b can be computed from the formula given previously. Now all we need to do is substitute any appropriate X-value for X, and compute Y_p, or the predicted Y-value. Consider the following example.

Example. Systolic blood-pressure measurements were taken on a number of human male subjects of selected ages, resulting in the paired data shown in Table 10.5. Compute the regression equation so that Y may be predicted from X. Calculate the predicted blood-pressure values for ages 35 and 50.

TABLE 10.5

Subject	X (Age in years)	Y (B.P. in mm Hg)
1	19	122
2	25	125
3	30	126
4	42	129
5	46	130
6	52	135
7	57	138
8	62	142
9	70	145

1. First, we need to calculate the following values from the data:

$$\sum X = 403 \qquad \sum Y = 1192$$

$$\sum X^2 = 20{,}463 \qquad \overline{Y} = 132.44$$

$$\frac{(\sum X)^2}{n} = 18{,}046 \qquad \sum XY = 54{,}461$$

$$\overline{X} = 44.78 \qquad \frac{\sum X \sum Y}{n} = 53{,}375$$

2. Now we may proceed to compute b, the slope of the regression equation, as

$$b = \frac{\sum XY - (\sum X \sum Y/n)}{\sum X^2 - [(\sum X)^2/n]} = \frac{54{,}461 - 53{,}375}{20{,}463 - 18{,}046}$$

$$= \frac{1086}{2417} = 0.45.$$

3. Now we may make the appropriate substitutions in the regression equation as follows:

$$Y_p = \overline{Y} + b(X - \overline{X})$$

$$= 132.44 + 0.45(X - 44.78).$$

Therefore, for age 35, the predicted blood pressure would be

$$Y_p = 132.44 + 0.45(35 - 44.78) = 132.44 + 0.45(-9.78)$$

$$= 132.44 - 4.40 = 128.04,$$

and for age 50,

$$Y_p = 132.44 + 0.45(50 - 44.78) = 132.44 + 0.45(5.22)$$

$$= 132.44 + 2.35 = 134.79.$$

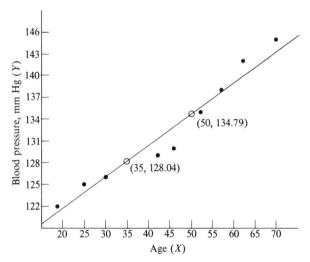

Fig. 10.7 Line of best fit based on Y-values predicted from ages 35 and 50.

Figure 10.7 shows a series of nine points representing the magnitudes of the nine X and Y pairs. By plotting two additional points based on the coordinates (35, 128.04) and (50, 134.79), and drawing a line through these points, we obtain the regression line, or "line of best fit." It will be recalled that this regression line contains all the *predicted* Y-values. Note that the slope of this line is positive; this is to be expected from the positive value of b in the preceding example.

Thus it may be seen that by using the regression equation, we have established the nature of the relationship between age and blood pressure, and we may also use this equation to predict the blood-pressure value associated with any specific age.

10.8 VARIANCE OF THE Y-VALUES
AROUND THE REGRESSION LINE

It must by now be obvious that when two variables are imperfectly correlated, the regression line is *estimated* from the data, and predicted Y-values cannot be considered precise. Also, the lower the correlation, the less precise any estimate of Y must be.

We therefore need to provide a means of estimating actual Y from Y_p with specified confidence limits. This involves computing the variance of Y around the regression line of Y_p's, using the familiar variance relationship

$$S_{YX}^2 = \frac{\Sigma(Y - Y_p)^2}{n - 2}.$$

Note that in computing S_{YX}^2, two degrees of freedom are lost.

As is often the case, the harmless looking formula above is the most difficult to use in actual practice, particularly when working with large amounts of data. Algebraic manipulation yields a formula that, though it appears more difficult, is more practical:

$$S_{YX}^2 = \frac{(\Sigma Y^2 - (\Sigma Y)^2/n) - b(\Sigma XY - \Sigma X \Sigma Y/n)}{n - 2}.$$

Taking the square root of S_{YX}^2 yields S_{YX}, the standard deviation of Y's around the regression line. Following the same procedure used in computing the standard error of the mean, we obtain the standard error of Y as:

$$S_{\bar{Y}} = \frac{S_{YX}}{\sqrt{n}}.$$

Now we can use the familiar estimation formula

$$Y_p \pm S_{\bar{Y}}(t),$$

where t is at the appropriate level (0.05 or 0.01) with $(n - 2)$ degrees of freedom.

Unfortunately, this procedure is reliable for estimating Y-values from X only when X is in the vicinity of $\bar{X}$. As we go farther away from $\bar{X}$ the confidence interval becomes larger, as illustrated by the biconcave dotted line in Fig. 10.8. Therefore, when estimating Y-values from X-values that are some distance from $\bar{X}$, a correction formula should be applied when calculating the standard error. This may be done as follows:

$$S_{\bar{Y}}^2(\text{corrected}) = S_{YX}^2 \left[\frac{1}{n} + \frac{(X - \bar{X})^2}{\Sigma(X - \bar{X})^2} \right],$$

where the desired X-value is substituted in the expression, $(X - \bar{X})^2$.

Example. Referring to the example involving age vs. blood pressure, given in Section 10.6, suppose that we wish to estimate the actual Y, from Y_p with 95 percent confidence limits based on an age of 65.

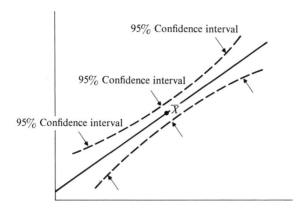

Fig. 10.8 Increasing confidence interval of Y as X deviates farther from $\overline{X}$.

1. Using the regression equation computed in Section 10.6, we calculate Y_p as follows:

$$\overline{Y} + b(X - \overline{X}) = 132.44 + 0.45(65 - 44.78)$$

$$= 132.44 + 0.45(20.22) = 132.44 + 9.10 = 141.54.$$

Thus, we would predict from our data that an age of 65 would be associated with a blood pressure of 141.54 mm Hg.

2. Now we need to compute the variance of the actual Y-values around the regression line, or S^2_{YX}. To do this, we need two additional pieces of data, $\sum Y^2$ and $(\sum Y)^2/n$. Calculating these, we obtain

$$\sum Y^2 = 158,384, \qquad \frac{(\sum Y)^2}{n} = 157,874.$$

Then

$$S^2_{YX} = \frac{(\sum Y^2 - (\sum Y)^2/N) - b(\sum XY - \sum X \sum Y/N)}{n - 2}$$

$$= \frac{(158,384 - 157,874) - 0.45(54,461 - 53,375)}{7}$$

$$= \frac{510 - 0.45(1086)}{7} = \frac{510 - 488.70}{7} = 3.04.$$

Then

$$S_{YX} = \sqrt{3.04} = 1.74,$$

and

$$S_{\overline{Y}} = \frac{1.74}{\sqrt{9}} = 0.58.$$

If our X-value were near the mean, we could proceed to substitute $S_{\overline{Y}} = 0.58$ in the estimation equation; but since 65 is a considerable distance from $\overline{X}$, we should compute the corrected $S_{\overline{Y}}$ as

$$S_{\overline{Y}}^2 \text{(corrected)} = S_{YX}^2\left[\frac{1}{n} + \frac{(X - \overline{X})^2}{\sum(X - \overline{X})^2}\right]$$

$$= 3.04\left[\frac{1}{9} + \frac{(65 - 44.78)}{2417}\right]$$

$$= 3.04(0.28) = 0.85.$$

Then

$$S_{\overline{Y}} = \sqrt{S_{\overline{Y}}^2} = \sqrt{0.85} = 0.92.$$

Note that the corrected $S_{\overline{Y}}$ is larger than the 0.58 value obtained as the uncorrected $S_{\overline{Y}}$, and will therefore contribute to a larger confidence interval.

3. Looking up the appropriate two-tailed t-value in the 0.05 column with $n - 2$, or 7 degrees of freedom, we obtain $t = 2.36$. Now we may use the following estimation formula to estimate true Y:

$$Y_p \pm S_{\overline{Y}}(t) = 141.54 \pm 0.92(2.36) = 141.54 \pm 2.17.$$

4. We may therefore state that, given an age of 65, we would estimate with 95 percent confidence limits that the associated blood-pressure value would be in the interval 139.37 to 143.71 mm Hg.

10.9 TESTING THE REGRESSION COEFFICIENT FOR SIGNIFICANCE

Like the correlation coefficient, the regression coefficient (slope) is derived from a sample, and it is necessary to test for the possibility that the sample regression coefficient, or slope, (b) is only a chance deviation from a true population regression coefficient (β) of zero. Regression analysis therefore involves testing the null hypothesis

$$H_0: \beta = 0,$$

where β represents the population regression coefficient. This can be done quite simply by using the following formula:

$$t = \frac{b - 0}{\sqrt{S_{YX}^2/[\sum(X - \overline{X})^2]}}.$$

Substituting the values derived from our blood-pressure problem data, we have the following:

$$t = \frac{0.45 - 0}{\sqrt{3.04/2417}} = 12.85.$$

Since we have used two parameter estimates in our computations (X, Y), we lose two degrees of freedom. We therefore enter the t-table at seven degrees of freedom and find that the t-value 12.85 is statistically significant beyond the 0.001 level. We therefore reject the null hypothesis, $H_0: \beta = 0$, and we may conclude that the population regression coefficient is statistically significantly different from zero.

10.10 NONLINEAR RELATIONSHIPS

In this chapter we have assumed that the relationships involving correlation and regression were linear. In other words, a scatter diagram resulting from Y plotted against X would be a straight line, or at least a roughly approximate straight line. This assumption is necessary if we are to be justified in applying the regression equation, which is itself a form of the basic linear equation, $Y = a + bX$.

In practice, biological data do not always take a linear form. The general pattern of population growth, for example, is represented by a curve similar to Fig. 10.9. Note that as time increases arithmetically, the population N increases geometrically until limiting factors such as food, oxygen, etc., begin to slow it down. This type of curve is rather common in biological work, and since we cannot apply the regression equation to a curve that is not linear, we need to apply a transformation that will literally straighten it out.

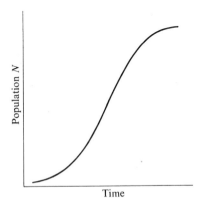

Fig. 10.9 General pattern of population growth.

This is easily accomplished by substituting the appropriate logarithm value for each value of the data showing nonlinear change. The following example will demonstrate the use of this transformation procedure.

Example. The data shown in Table 10.6 were obtained from periodic observations made on the growth of a population of yeast cells; the counts were taken every two hours.

TABLE 10.6

Hours	Number of cells
2	19
4	37
6	72
8	142
10	295
12	584
14	995

1. Figure 10.10 shows the nonlinear curve that results when the cell number is plotted against time. Note that cell number increases exponentially while time increases arithmetically.

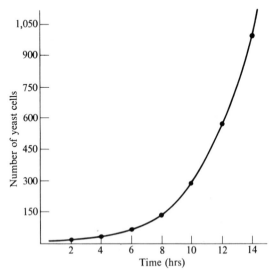

Fig. 10.10 Nonlinear curve resulting from number of yeast cells plotted against time.

2. In order to "straighten out" the curve, we will substitute the common logarithm value for each cell count, yielding the values shown in Table 10.7.

TABLE 10.7

Count	Log
19	1.2788
37	1.5682
72	1.8573
142	2.1523
295	2.4698
584	2.7664
995	2.9978

3. Figure 10.11 shows the curve resulting from plotting the log values of the cell counts against time. This new curve is linear in nature, permitting the application of the principles of linear regression considered in previous sections.

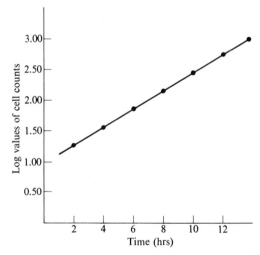

Fig. 10.11 Log values of yeast-cell counts plotted against time.

PROBLEMS

10.1 A sample of twelve leaves were randomly collected from a tree and the length and width of each leaf were measured to the nearest millimeter. Referring to the data given in Table 10.8 (see Sections 10.3 through 10.6),

a) Compute the sample correlation coefficient.
b) Test the sample correlation coefficient for statistical significance.
c) Estimate the upper and lower confidence limits for ρ and state your conclusions concerning length and width in terms of the coefficient of determination.

TABLE 10.8

Leaf	Width	Length
1	35	55
2	21	44
3	25	46
4	35	60
5	26	55
6	40	57
7	35	64
8	40	68
9	25	51
10	42	61
11	23	46
12	25	44

10.2 Given the bivariate distribution shown in Table 10.9, in which both X and Y are random variables (see Sections 10.3 through 10.6),

a) Compute the correlation coefficient;
b) Test the sample correlation coefficient for statistical significance;
c) If r is statistically significant, estimate the upper and lower limits of ρ and state your conclusions concerning the variability in X that is associated with the variability in Y.

TABLE 10.9

Number	X	Y
1	50	20
2	54	19
3	36	23
4	63	18
5	53	20
6	49	21
7	52	18
8	58	17
9	46	16
10	45	25

10.3 A bivariate distribution with $n = 15$ yields a sample correlation coefficient of 0.75. (See Sections 10.3 through 10.6.)

a) Test the sample r for statistical significance;

b) Estimate the upper and lower limits of ρ with 95 percent confidence.

c) At each of the limits found in (b), determine the percentage of variability in one variable that appears to be associated with the other.

10.4 Repeat the steps listed in Problem 10.3, but this time assume a correlation coefficient of 0.75 derived from a sample of $n = 100$. (See Sections 10.3 through 10.6.)

10.5 An experimenter wished to determine the relationship between temperature and heartbeat rate in the common grass frog, *Rana pipiens*. The temperature was manipulated in 2-degree increments ranging from 2°C to 18°C, with heart-beat rates recorded at each interval. Analyze the data given in Table 10.10, in terms of the following (see Sections 10.7 and 10.8):

a) Compute the regression coefficient (slope).

b) Construct a "line of best fit" to the data.

c) Test the regression coefficient for statistical significance.

d) Compute the upper and lower 95-percent confidence limits for the heart rate expected in *Rana pipiens* at 9°C.

TABLE 10.10

Recording number	Temperature (° Celsius)	Heartbeat (per minute)
1	2	5
2	4	11
3	6	11
4	8	14
5	10	22
6	12	23
7	14	32
8	16	29
9	18	32

10.6 An experiment was performed to determine the relationship between age and heart rate (in beats per minute) in human females ranging from one to fifteen years. Analyze the data in Table 10.11 in terms of the following (see Sections 10.7 and 10.8):

a) Compute the regression coefficient (slope).

b) Construct a "line of best fit" to the data.

c) Test the regression coefficient for statistical significance.

d) Compute the upper and lower 95-percent confidence limits for the heart rate expected in a randomly selected ten-year-old female child.

TABLE 10.11

Recording number	Age	Heart rate
1	1	111
2	2	108
3	3	108
4	4	102
5	5	99
6	6	92
7	7	93
8	8	88
9	9	90
10	10	90
11	11	88
12	12	86
13	13	84
14	14	83
15	15	83

CHAPTER ELEVEN

Analysis of Covariance

11.1 INTRODUCTION

In Chapter 9 we applied the analysis of variance to multiple-group designs, and Chapter 10 included the concept of regression analysis. In this chapter, we will show an analysis of data by a technique which combines analysis of variance with the principles of regression analysis. This is called *analysis of covariance*, and it is a powerful statistical tool for attacking problems found in many areas of biological research.

Let us begin by posing a very simple problem. Suppose that we wish to compare two methods of teaching statistics in terms of performance on an examination. Two classes, randomly selected, will be taught by methods A and B, respectively. At the end of the course, we will compare the final examination scores achieved by the class taught by method A with those achieved by the class taught by method B. The results could then be analyzed by the usual method, either with a *t*-test or with analysis of variance.

Now, any conclusions drawn from this analysis will necessarily assume that the two groups were pretty much alike to begin with, at least as far as knowledge of statistics is concerned! In other words, if the results are to be considered valid, the students in each of the two classes must start the course with similar backgrounds. To carry this concept to the ultimate, what conclusions could be drawn if the "method A" group were composed of statisticians and the "method B" group contained only students who had never heard of the word?

Consider the hypothetical sets of scores obtained on the examination, as shown in Table 11.1.

No statistical analysis is needed to see that group B obviously fared much better on the examination, and it would seem that method B has a great deal going for it.

TABLE 11.1

Group A		Group B	
Student	Score	Student	Score
1	70	1	92
2	72	2	84
3	68	3	80
4	81	4	95
5	78	5	90

But now, suppose that we had given the same examination as a pretest when the course was just beginning, and further suppose the results listed in Table 11.2.

TABLE 11.2

	Group A			Group B	
Student	Pretest (X_A)	Post-test (Y_A)	Student	Pretest (X_B)	Post-test (Y_B)
1	45	70	1	60	92
2	40	72	2	61	84
3	38	68	3	58	80
4	54	81	4	73	95
5	42	78	5	71	90

From this somewhat extreme example, it is now clear that comparing group A with group B is like putting the author in the ring with the heavyweight champion of the world. It is therefore apparent that no conclusion can be drawn concerning the comparative efficiency of the two methods of teaching statistics.

This same concept can easily be applied to such variables as blood pressure, cholesterol content, plot production, and a host of situations found in biological experimentation. If, for example, we wish to compare fertilizers by comparing the yields of treated plots, it is apparent that the differential effects of *previous* fertilizer treatments must be considered. In general, we could more accurately evaluate treatment effects of various kinds if we could take into consideration the value of the variable found in each subject *before* the treatment is applied. This would be especially useful when working with biological data, where values are often questionable.

The analysis of covariance makes use of information in the form of pretreatment data in order to compensate for the effects of prior treatment or the pretreatment condition. This pretreatment measurement is usually called the concomitant variable, or control variable. It may be seen that such control variables are essentially *predictors* of post-treatment measurements.

11.2 RATIONALE UNDERLYING ANALYSIS OF COVARIANCE

Suppose that we wish to compare the effects of two different diets on the weights of mice. We shall use five mice per diet, keeping the number small so that the arithmetic will be less tedious.

We could, of course, simply randomize the assignment of the ten mice to the diets, and statistically compare their weights after a specific time period, using the unmatched t-test or analysis of variance. Suppose that we were to do this, and further suppose that we obtain the set of post-treatment weights in grams shown in Table 11.3.

TABLE 11.3

Diet A	Diet B
80	54
51	78
78	59
81	61
72	78

Now, it would be useful to know what each of these subjects weighed before the diet treatment, since the final weight of each mouse will quite obviously be affected in part by what the mouse weighed in the first place. Suppose that we had kept just such a record, and we can now add this pretreatment weight to the data. Our complete set of data would then appear as shown in Table 11.4, where the X's are the pretreatment weights, or *control* variable, and the Y's are the post-treatment weights.

It is fairly obvious that an association exists between pretreatment weights (X) and post-treatment weights (Y). This is confirmed by the scatter diagram shown in Fig. 11.1.

It is therefore apparent that the postdiet weights are influenced by initial weights as well as by diet effects. It is further apparent that a more accurate comparison might be made if we were to adjust the postdiet weights according to the prediet values. In other words, we could adjust

TABLE 11.4

Diet A		Diet B	
X_A	Y_A	X_B	Y_B
60	80	42	54
38	51	52	78
54	78	45	59
55	81	43	61
50	72	50	78
$\bar{X}_A = 51.40$	$\bar{Y}_A = 72.40$	$\bar{X}_B = 46.40$	$\bar{Y}_B = 66.00$

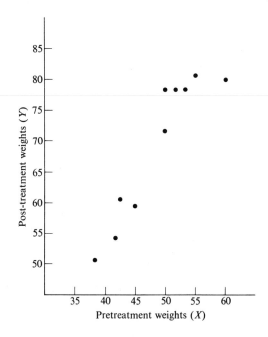

Fig. 11.1 Scatter diagram showing correlation between pre- and post-treatment weights.

the Y-values on the basis of the initial advantage or disadvantage possessed by the individual animal.

This could be done by computing a regression coefficient based on the relationship between the X's and Y's in the total distribution. Computing this regression coefficient by the method described in Chapter 10, we obtain $b = 1.61$. Now, computing the mean of all ten X-values yields 48.90. We may now determine the initial advantage or disadvantage in each case by multiplying the regression coefficient, 1.61, by the difference between each prediet weight and the mean of all prediet weights, 48.90. Where the initial weight is *above* the mean, we will subtract the initial advantage from the associated postdiet weight. Where the initial weight is *less* than the mean of all initial weights, we will add the initial disadvantage to the associated postdiet value. In this way we may *adjust* the Y-values for initial weights; in other words, we tend to negate pretreatment advantages or disadvantages that might be present.

The results of this adjusting procedure are listed in Table 11.5.

TABLE 11.5

Diet A X_A	Y_A	$Y_A adj.$
$60 - 48.90 \times 1.61 = 17.87$	80	62.13
$38 - 48.90 \times 1.61 = 17.55$	51	68.55
$54 - 48.90 \times 1.61 = 8.21$	78	69.79
$55 - 48.90 \times 1.61 = 9.82$	81	71.18
$50 - 48.90 \times 1.61 = 1.77$	72	70.23

Diet B X_B	Y_B	$Y_B adj.$
$42 - 48.90 \times 1.61 = 11.11$	54	65.11
$52 - 48.90 \times 1.61 = 4.99$	78	73.01
$45 - 48.90 \times 1.61 = 6.28$	59	65.28
$43 - 48.90 \times 1.61 = 9.50$	61	70.50
$50 - 48.90 \times 1.61 = 1.77$	78	76.23

We could now perform analysis of variance on the *adjusted* Y-values, and the resulting F-value could be assumed to be free from bias due to initial weights.

As is so often the case in statistics, this "simple" approach is obviously tedious, even with the unrealistically small amount of data used in the illustration. As usual, we will look for a method that will accomplish the same end result, but with less time and effort.

This time, we shall begin by plotting a regression line based on the same bivariate data shown in the preceding illustration. A review of Chapter 10 shows that we can predict the Y-value (Y_p) that will be associated with a specific X-value by using the regression equation,

$$Y_p = \bar{Y} + b(X - \bar{X}).$$

Figure 11.2 shows a regression line formed by the Y_p-values which were predicted from the weights taken before the treatment. The Y-values, or *actual* post-treatment weights, are also shown, together with the distance of each actual Y-value from its associated Y_p, or *predicted* value.

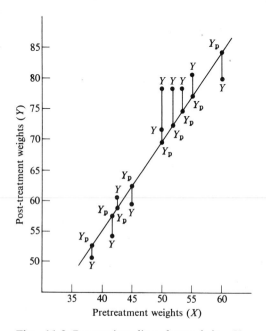

Fig. 11.2 Regression line formed by Y_p-values predicted from pretreatment weights.

Now we come to the essential rationale underlying the technique of covariance. The total variability in Y consists of two elements: (1) the variability in Y due to the variability in X, and (2) the variability in Y that is due to the components involving treatment and experimental error. It may be seen that if we can remove that variability in Y due to the variability in X, we could then analyze the resulting data in terms of treatment and error. This data would be uncontaminated, as it were, by the influence of the pretreatment values.

Now, expressing this idea in the form of an equation, we have

$$\sum(Y - \bar{Y})^2 = \sum(Y_p - \bar{Y})^2 + \sum(Y - Y_p)^2,$$

which, symbolically, states that the total variability in Y is equal to the variability in Y predicted from the variability in X, plus the variability in Y that is *not* predicted from X. Since it is this latter variability which is of interest, we shall rearrange the equation so that we have

$$\sum(Y - Y_\mathrm{p})^2 = \sum(Y - \overline{Y})^2 - \sum(Y_\mathrm{p} - \overline{Y})^2.$$

Since the expression $\sum(Y_\mathrm{p} - \overline{Y})^2$ represents the variability in Y due to pretreatment, or X-values, and since it may be predicted from those X-values, we simply need to assess this predicted variability and remove it from the total variability in Y. To do this, we need to recall that the regression coefficient or slope of the line is computed as

$$b = \frac{\sum(X - \overline{X})(Y - \overline{Y})}{\sum(X - \overline{X})^2}$$

or, in deviation symbols,

$$b = \frac{\sum xy}{\sum x^2}.$$

Now, since

$$(Y_\mathrm{p} - \overline{Y}) = b(X - \overline{X}),$$

a little algebraic monkey business leads to the following equation:

$$\sum(Y - Y_\mathrm{p})^2 = \sum(Y - \overline{Y})^2 - b\sum xy,$$

where the expression $b\sum xy$ has been substituted as an identity to $\sum(Y_\mathrm{p} - \overline{Y})^2$.

This is the critical equation of covariance, and is used to compute the variability in Y due to X, so that it may be removed from the total variability in Y. This leaves $\sum(Y - Y_\mathrm{p})^2$, which is the variability in Y due to the influence of treatment and error.

It will be recalled that in performing the analysis of variance, we separated the variability of Y into two components: (1) the variability due to treatment effect, and (2) the variability due to experimental error, or sampling error. Our next step in covariance, then, is to partition the remaining variability in Y into two similar components. The significance of the treatment effect can then be assessed in the usual manner by the F-ratio.

11.3 SINGLE-CLASSIFICATION ANALYSIS OF COVARIANCE

In the preceding section, we discussed the theoretical basis underlying covariance. Now we will get down to practical matters involving the application of this technique. As an illustration, we will work through a single-classification analysis in a step-by-step fashion, using an extension of the mouse diet problem.

Example. An experiment using two randomly assigned groups of mice was performed to determine the comparative effects of diet A and diet B. The pretreatment weights were recorded and the final weights were determined after a specified treatment time, resulting in the data listed in Table 11.6.

TABLE 11.6

| | Diet A | | | Diet B | |
| | Prediet | Postdiet | | Prediet | Postdiet |
Animal	X_A	Y_A	Animal	X_B	Y_B
1	60	80	1	58	81
2	55	81	2	46	58
3	54	78	3	50	75
4	50	72	4	39	60
5	38	51	5	41	59
6	42	54	6	45	60
7	50	78	7	42	58
8	45	59	8	55	72
9	43	61	9	52	75
10	52	78	10	45	57
	$\bar{X}_A = 48.90$ $\bar{Y}_A = 69.20$			$\bar{X}_B = 47.30$ $\bar{Y}_B = 65.50$	

1. The first step in the analysis involves the computation of several "pieces" of data. This may appear somewhat tedious at first, but with a desk calculator it may be done rather quickly. The data needed, with their numerical values, are:

$$\sum XY = 65{,}943 \qquad \sum X = 962 \qquad \sum Y = 1347$$
$$\sum X^2 = 47{,}052 \qquad \sum X_A = 489 \qquad \sum Y_A = 692$$
$$\sum Y^2 = 92{,}769 \qquad \sum X_B = 473 \qquad \sum Y_B = 655$$
$$n_T = 20 \qquad\qquad n_A = 10 \qquad\qquad n_B = 10$$

Note that the absence of a subscript indicates the sum of *all* X's, Y's, or XY cross products. The subscript is used to indicate the specific treatment with which the value is concerned. For example, the sum of all X's involved with diet A is symbolized by $\sum X_A$.

2. Next, we partition the total SS of X-values into SS for diet and SS for within, or error:

$$SS_{Total} = \sum X^2 - \frac{(\sum X)^2}{n_T} = 47,052 - \frac{(962)^2}{20}$$

$$= 47,052 - 46,272 = 780,$$

$$SS_{Diet} = \frac{(\sum X_A)^2}{n_A} + \frac{(\sum X_B)^2}{n_B} - \frac{(\sum X)^2}{n_T}$$

$$= \frac{(489)^2}{10} + \frac{(473)^2}{10} - \frac{(962)^2}{20}$$

$$= 46,285 - 46,272 = 13,$$

$$SS_{Within} = SS_{Total} - SS_{Diet} = 780 - 13 = 767.$$

3. Next, we partition the total SS for the Y-variable into SS for diet and SS for within:

$$SS_{Total} = \sum Y^2 - \frac{(\sum Y)^2}{n_T} = 92,769 - \frac{(1347)^2}{20}$$

$$= 92,769 - 90,720 = 2049,$$

$$SS_{Diet} = \frac{(\sum Y_A)^2}{n_A} + \frac{(\sum Y_B)^2}{n_B} - \frac{(\sum Y)^2}{n_T}$$

$$= \frac{(692)^2}{10} + \frac{(655)^2}{10} - \frac{(1347)^2}{20}$$

$$= 90,789 - 90,720 = 69,$$

$$SS_{Within} = SS_{Total} - SS_{Diet} = 2049 - 69 = 1980.$$

4. Now we partition the SS for cross products (XY) into SS for diet and SS for within:

$$SS_{Total} = \sum XY - \frac{\sum X \sum Y}{n}$$

$$= 65,943 - \frac{(962)(1347)}{20} = 1152,$$

$$SS_{Diet} = \frac{(\sum X_A)(\sum Y_A)}{n_A} + \frac{(\sum X_B)(\sum Y_B)}{n_B} - \frac{\sum X \sum Y}{n_T}$$

$$= \frac{(489)(692)}{10} + \frac{(473)(655)}{10} - \frac{(962)(1347)}{20} = 30,$$

$$SS_{Within} = SS_{Total} - SS_{Diet} = 1152 - 30 = 1122.$$

5. Now we set up Table 11.7 containing the computed SS as follows:

TABLE 11.7

Source	SS(X)	SS(Y)	SS(XY)
Total	780	2049	1152
Diet	13	69	30
Within	767	1980	1122

6. Next, we add the within SS in each case to the diet SS. This yields "within plus diet" values:

SS(X)	SS(Y)	SS(XY)
780	2049	1152

7. Now we compute the regression coefficient for the "within plus diet" values. This is done by the use of the formula

$$b = \frac{\sum xy}{\sum x^2},$$

where "$\sum xy$" is another way of expressing SS(XY), and "$\sum x^2$" is another symbol for SS(X). Thus

$$b = \frac{1152}{780} = 1.48.$$

8. Now we apply the basic covariance equation. This allows the removal of the variability in Y that is due to, and predicted from, the variability in X. Thus

$$\sum(Y - Y_{\mathrm{p}})^2 = \sum(Y - \bar{Y})^2 - b\sum xy$$

$$= 2049 - (1.48)(1152)$$

$$= 2049 - 1705 = 344.$$

9. Next, we perform the same operation for the "within" sum of squares in order to assess the experimental error when the variability due to X is removed. Thus

$$b = \frac{\sum xy}{\sum x^2} = \frac{1122}{767} = 1.46$$

and

$$\sum(Y - Y_{\mathrm{p}})^2 = \sum(Y - \bar{Y})^2 - b\sum xy$$

$$= 1980 - (1.46)(1122)$$

$$= 1980 - 1638 = 342.$$

10. To determine degrees of freedom associated with each source, we begin by subtracting one from the total, leaving 19. Now we subtract one degree of freedom associated with the control variable (X) and one degree of freedom associated with the main effect from 19. This yields 17 as the degrees of freedom associated with the within, or error effect. Thus we have the results shown in Table 11.8.

TABLE 11.8

Source	d.f.	SS	MS
Diet	1	2	2.00
Within	17	342	20.12
Within + diet		344	

Note that the SS due to diet alone has been found by subtracting the residual SS for "within" from the SS for "within plus diet."

11. Computing the F-value in the usual way, we have

$$F = \frac{\text{MS(Diet)}}{\text{MS(Within)}} = \frac{2.00}{20.12} = 0.099,$$

and since the obtained F is below unity, we automatically assume a lack of significance. Therefore, we must conclude that our experimental and statistical evidence does not support a contention that diet A and diet B produce significantly different effects in terms of final weights.

11.4 COVARIANCE APPLIED TO A FACTORIAL EXPERIMENT

In the previous section, analysis of covariance was applied to a single-factor design. In this section the covariance technique will be applied to a more complicated situation involving a factorial experiment. The principles related to factorial experiments that were discussed in Chapter 9 also apply here, and the similarity of the computational procedure to that of analysis of variance should again be noted.

Example. An investigator wishes to test three special diets in terms of their differential effects on reducing cholesterol content. In addition, he wishes to test the effect of a drug on cholesterol content, and finally, he is interested in the presence of possible interaction effects between the diet and

TABLE 11.9

	Diet A		Diet B		Diet C		
	X_{AP}	Y_{AP}	X_{BP}	Y_{BP}	X_{CP}	Y_{CP}	
Placebo	180	145	190	150	190	150	
	195	150	185	155	180	140	
	205	160	200	150	190	160	
	200	155	190	150	185	150	
	195	150	185	140	185	150	
	$\bar{X}_{AP} = 195.00$ $\bar{Y}_{AP} = 152.00$		$\bar{X}_{BP} = 190.00$ $\bar{Y}_{BP} = 149.00$		$\bar{X}_{CP} = 186.00$ $\bar{Y}_{CP} = 150.00$		$\bar{X}_P = 190.33$ $\bar{Y}_P = 150.33$
	X_{AD}	Y_{AD}	X_{BD}	Y_{BD}	X_{CD}	Y_{CD}	
Drug	195	145	200	160	185	145	
	210	155	195	150	195	155	
	195	140	190	140	205	160	
	200	160	195	150	175	140	
	180	140	190	155	185	140	
	$\bar{X}_{AD} = 197.60$ $\bar{Y}_{AD} = 148.00$		$\bar{X}_{BD} = 194.00$ $\bar{Y}_{BD} = 151.00$		$\bar{X}_{CD} = 189.00$ $\bar{Y}_{CD} = 148.00$		$\bar{X}_D = 193.53$ $\bar{Y}_D = 149.67$
	$\bar{X}_A = 196.80$ $\bar{Y}_A = 150.00$		$\bar{X}_B = 192.00$ $\bar{Y}_B = 150.00$		$\bar{X}_C = 187.50$ $\bar{Y}_C = 149.00$		

the drug. He therefore designs a factorial experiment, where the three diets are the levels of one factor, and the drug and a placebo are the levels of the other factor. Thirty adult males are randomly assigned in groups of five to each of the combinations of factors. Subjects were chosen from a limited age range to minimize the effect of age on cholesterol content. The cholesterol content of all subjects was measured before and after treatment, and Table 11.9 shows the general design along with the raw data, where cholesterol content is measured in grams/100 ml serum. Note that the subscripts are descriptive in terms of the factor or combination of factors involved.*

1. Before starting the actual computations, we shall first code the raw data by subtracting 140 from each value. This makes the work easier, even when using a calculator. Remember, subtracting a constant from each member of a distribution in no way changes the variance of that distribution; therefore we can base our computations on the coded data listed in Table 11.10.

TABLE 11.10

Diet A		Diet B		Diet C	
X_{AP}	Y_{AP}	X_{BP}	Y_{BP}	X_{CP}	Y_{CP}
40	5	50	10	50	10
55	10	45	15	40	0
65	20	60	10	50	20
60	15	50	10	45	10
55	10	45	0	45	10
X_{AD}	Y_{AD}	X_{BD}	Y_{BD}	X_{CD}	Y_{CD}
55	5	60	20	45	5
70	15	55	10	55	15
55	0	50	0	65	20
60	20	55	10	35	0
48	0	50	15	45	0

2. Again, we need to begin by computing several "pieces" of data, as shown in Table 11.11.

3. Next, we partition the SS for the X-variable into the SS for diet, the SS for drug, the SS for interaction, and the SS for within, or error. Thus

---

* We have been using a lower-case subscript p for predicted values, so we shall use capital subscript P to identify values related to the placebo.

$$SS_{Total} = \sum X^2 - \frac{(\sum X)^2}{n_T} = 82{,}804 - \frac{(1558)^2}{30}$$

$$= 82{,}804 - 80{,}912 = 1892,$$

$$SS_{Diet} = \frac{(\sum X_A)^2}{n_A} + \frac{(\sum X_B)^2}{n_b} + \frac{(\sum X_C)^2}{n_C} - \frac{(\sum X)^2}{n_T}$$

$$= \frac{(563)^2}{10} + \frac{(520)^2}{10} + \frac{(475)^2}{10} - \frac{(1558)^2}{30}$$

$$= 81{,}299 - 80{,}912 = 387,$$

$$SS_{Drug} = \frac{(\sum X_D)^2}{n_D} + \frac{(\sum X_P)^2}{n_P} - \frac{(\sum X)^2}{n_T}$$

$$= \frac{(803)^2}{15} + \frac{(755)^2}{15} - \frac{(1558)^2}{30}$$

$$= 80{,}989 - 80{,}912 = 77,$$

$$SS_{Interaction} = \frac{(\sum X_{AP})^2}{n_{AP}} + \frac{(\sum X_{BP})^2}{n_{BP}} + \frac{(\sum X_{CP})^2}{n_{CP}}$$

$$+ \frac{(\sum X_{AD})^2}{n_{AD}} + \frac{(\sum X_{BD})^2}{n_{BD}} + \frac{(\sum X_{CD})^2}{n_{CD}}$$

$$- \frac{(\sum X)^2}{n_T} - (SS_{Drug} + SS_{Diet})$$

$$= \frac{(275)^2}{5} + \frac{(250)^2}{5} - \frac{(230)^2}{5} + \frac{(288)^2}{5}$$

$$+ \frac{(270)^2}{5} + \frac{(245)^2}{5} - \frac{(1558)^2}{30} - (77 + 387)$$

$$= 81{,}379 - 80{,}912 - (77 + 387) = 3,$$

$$SS_{Within} = SS_{Total} - (SS_{Drug} + SS_{Diet} + SS_{Interaction})$$

$$= 1892 - (77 + 387 + 3) = 1425.$$

4.   Next, we partition the SS for the Y-variable into the SS for diet, the SS for drug, the SS for interaction, and the SS for within, or error. Thus

$$SS_{Total} = \sum Y^2 - \frac{(\sum Y)^2}{n_T} = 4200 - \frac{(290)^2}{30}$$

$$= 4200 - 2804 = 1396,$$

TABLE 11.11

$\sum XY = 16{,}100$	$\sum X = 1558$	$\sum Y = 290$
$\sum X^2 = 82{,}804$	$\sum X_A = 563$	$\sum Y_A = 100$
$\sum Y^2 = 4200$	$\sum X_B = 520$	$\sum Y_B = 100$
$n_T = 30$	$\sum X_C = 475$	$\sum Y_C = 90$
	$\sum X_D = 803$	$\sum Y_D = 135$
	$\sum X_P = 755$	$\sum Y_P = 155$
	$\sum X_{AP} = 275$	$\sum Y_{AP} = 60$
	$\sum X_{BP} = 250$	$\sum Y_{BP} = 45$
	$\sum X_{CP} = 230$	$\sum Y_{CP} = 50$
	$\sum X_{AD} = 288$	$\sum Y_{AD} = 40$
	$\sum X_{BD} = 270$	$\sum Y_{BD} = 55$
	$\sum X_{CD} = 245$	$\sum Y_{CD} = 40$

$$
\begin{aligned}
\mathrm{SS_{Diet}} &= \frac{(\sum Y_A)^2}{n_A} + \frac{(\sum Y_B)^2}{n_B} + \frac{(\sum Y_C)^2}{n_C} - \frac{(\sum Y)^2}{n_T} \\
&= \frac{(100)^2}{10} + \frac{(100)^2}{10} + \frac{(90)^2}{10} - \frac{(290)^2}{30} \\
&= 2810 - 2804 = 6,
\end{aligned}
$$

$$
\begin{aligned}
\mathrm{SS_{Drug}} &= \frac{(\sum Y_D)^2}{n_D} + \frac{(\sum Y_P)^2}{n_P} - \frac{(\sum Y)^2}{n_T} \\
&= \frac{(135)^2}{15} + \frac{(155)^2}{15} - \frac{(290)^2}{30} \\
&= 2817 - 2804 = 13,
\end{aligned}
$$

$$
\begin{aligned}
\mathrm{SS_{Interaction}} &= \frac{(\sum Y_{AP})^2}{n_{AP}} + \frac{(\sum Y_{BP})^2}{n_{BP}} + \frac{(\sum Y_{CP})^2}{n_{CP}} \\
&\quad + \frac{(\sum Y_{AD})^2}{n_{AD}} + \frac{(\sum Y_{BD})^2}{n_{BD}} + \frac{(\sum Y_{CD})^2}{n_{CD}} \\
&\quad - \frac{(\sum Y)^2}{n_T} - (\mathrm{SS_{Drug}} + \mathrm{SS_{Diet}}) \\
&= \frac{(60)^2}{5} + \frac{(45)^2}{5} + \frac{(50)^2}{5} + \frac{(40)^2}{5} \\
&\quad + \frac{(55)^2}{5} + \frac{(40)^2}{5} - 2804 - (6 + 13) \\
&= 2870 - 2804 - (6 + 13) = 47,
\end{aligned}
$$

$$SS_{\text{Within}} = SS_{\text{Total}} - (SS_{\text{Drug}} + SS_{\text{Diet}} + SS_{\text{Interaction}})$$

$$= 1396 - (6 + 13 + 47) = 1330.$$

5.   Finally, we partition the SS for the cross products (XY) into the SS for diet, the SS for drug, the SS for interaction, and the SS for within, or error. Thus

$$SS_{\text{Total}} = \sum XY - \frac{\sum X \sum Y}{n_{\text{T}}} = 16{,}100 - \frac{(1558)(290)}{30}$$

$$= 16{,}100 - 15{,}061 = 1039,$$

$$SS_{\text{Diet}} = \frac{\sum X_{\text{A}} \sum Y_{\text{A}}}{n_{\text{A}}} + \frac{\sum X_{\text{B}} \sum Y_{\text{B}}}{n_{\text{B}}} + \frac{\sum X_{\text{C}} \sum Y_{\text{C}}}{n_{\text{C}}} - \frac{\sum X \sum Y}{n_{\text{T}}}$$

$$= \frac{(563)(100)}{100} + \frac{(520)(100)}{10} + \frac{(475)(90)}{10} - \frac{(1558)(290)}{30}$$

$$= 5630 + 5200 + 4266 - 15{,}061$$

$$= 15{,}096 - 15{,}061 = 35,$$

$$SS_{\text{Drug}} = \frac{\sum X_{\text{D}} \sum Y_{\text{D}}}{n_{\text{D}}} + \frac{\sum X_{\text{P}} \sum Y_{\text{P}}}{n_{\text{P}}} - \frac{\sum X \sum Y}{n_{\text{T}}}$$

$$= \frac{(803)(135)}{15} + \frac{(755)(155)}{15} - \frac{(1558)(290)}{30}$$

$$= 7227 + 7802 - 15{,}061$$

$$= 15{,}029 - 15{,}061 = -32,$$

$$SS_{\text{Interaction}} = \frac{\sum X_{\text{AP}} \sum Y_{\text{AP}}}{n_{\text{AP}}} + \frac{\sum X_{\text{BP}} \sum Y_{\text{BP}}}{n_{\text{BP}}} + \frac{\sum X_{\text{CP}} \sum Y_{\text{CP}}}{n_{\text{CP}}}$$

$$+ \frac{\sum X_{\text{AD}} \sum Y_{\text{AD}}}{n_{\text{AD}}} + \frac{\sum X_{\text{BD}} \sum Y_{\text{BD}}}{n_{\text{BD}}}$$

$$+ \frac{\sum X_{\text{CD}} \sum Y_{\text{CD}}}{n_{\text{CD}}} - \frac{\sum X \sum Y}{n_{\text{T}}}$$

$$- (SS_{\text{Diet}} + SS_{\text{Drug}})$$

$$= \frac{(275)(60)}{5} + \frac{(250)(45)}{5} + \frac{(230)(50)}{5}$$

$$+ \frac{(288)(40)}{5} + \frac{(270)(55)}{5} + \frac{(245)(40)}{5}$$

$$- \frac{(1558)(290)}{30} - (35 - 32)$$

$$= 15{,}084 - 15{,}061 - (35 - 32) = 20,$$

$$SS_{Within} = SS_{Total} - (SS_{Diet} + SS_{Drug} + SS_{Interaction})$$
$$= 1039 - (35 - 32 + 20) = 1016.$$

6. As before, we now set up a table (Table 11.12) containing the computed SS.

TABLE 11.12

Source	SS(X)	SS(Y)	SS(XY)
Total	1892	1396	1039
Diet	387	6	35
Drug	77	13	-32
Interaction	3	47	20
Within	1425	1330	1016

7. Next, we add the within SS in each case to the appropriate source, yielding the "within plus effect" values shown in Table 11.13.

TABLE 11.13

Source	SS(X)	SS(Y)	SS(XY)
Diet	1812	1336	1051
Drug	1502	1343	984
Interaction	1428	1377	1036

8. Next, referring to the "within plus" values, we compute the SS for residuals in the case of each effect. In other words, we remove from the variability in Y that component of variability that is due to and predicted from X. Thus

Diet

$$b = \frac{\sum xy}{\sum x^2} = \frac{1051}{1812} = 0.580,$$

$$\sum (Y - Y_p)^2 = \sum (Y - \overline{Y})^2 - b\sum xy = 1336 - (0.580)(1051)$$
$$= 1336 - 610 = 726.$$

Drug

$$b = \frac{\sum xy}{\sum x^2} = \frac{984}{1502} = 0.655,$$

$$\sum (Y - Y_p)^2 = \sum (Y - \overline{Y})^2 - b\sum xy = 1343 - (0.655)(984)$$
$$= 1343 - 645 = 698.$$

Interaction

$$b = \frac{\sum xy}{\sum x^2} = \frac{1036}{1428} = 0.725,$$

$$\sum(Y - Y_p)^2 = \sum(Y - \overline{Y})^2 - b\sum xy = 1377 - (0.725)(1036)$$
$$= 1377 - 751 = 626.$$

9. Computing the SS for residuals for the "within," or error component, we have

$$b = \frac{\sum xy}{\sum x^2} = \frac{1016}{1425} = 0.712,$$

$$\sum(Y - Y_p)^2 = \sum(Y - \overline{Y})^2 - b\sum xy = 1330 - (0.712)(1016)$$
$$= 1330 - 723 = 607.$$

10. Now we are ready to compute the mean squares in order to perform the appropriate F-tests. To obtain the degrees of freedom in each case, we begin by subtracting one from the total, or $30 - 1$, leaving 29. Since there are three diets, two degrees of freedom are associated with the diet effect. Since there are two drugs, one degree of freedom is associated with the drug effect. Multiplying 2 times 1 yields two degrees of freedom associated with interaction between the diet and the drug. Subtracting 2, 1, and 2 from 29 leaves 24. Then we subtract one degree of freedom associated with the control variable from 24, and we have 23 degrees of freedom associated with the within, or error component.

11. The analysis-of-variance table therefore appears as in Table 11.14 where the residual sum of squares for within has been subtracted from the residual sum of squares for each "within plus effect" found in step 8.

12. Computing the F-values, we have

$$F_{2,23}(\text{diet}) = \frac{\text{MS(diet)}}{\text{MS(within)}} = \frac{59.50}{26.39} = 2.25,$$

$$F_{1,23}(\text{drug}) = \frac{\text{MS(drug)}}{\text{MS(within)}} = \frac{91.00}{26.39} = 3.45,$$

$$F_{2,23}(\text{interaction}) = \frac{\text{MS(interaction)}}{\text{MS(within)}} = \frac{9.50}{26.39} = 0.36.$$

13. Consulting Table VII, we find that none of the F-values is significant at the 0.05 level. We must therefore conclude that no significant difference exists among diets A, B, and C, in terms of effect on choles-

TABLE 11.14

	d.f.	SS	MS
Diet	2	119	59.50
Drug	1	91	91.00
Interaction	2	19	9.50
Within	23	607	26.39

terol content. Similarly, we must conclude that the drug did not significantly lower the cholesterol content of the drug group below that of the placebo group. Finally, we have found no evidence of interaction between the diet and the drug.

11.5 SUMMARY

It may be seen from the foregoing material that covariance is a very useful and powerful statistical tool. It should also be noted that while the computations are admittedly somewhat lengthy, they are not as difficult as they at first appear, providing the procedure is worked through in a step-by-step fashion.

The basic similarity to analysis of variance should be noted. In fact, the assumptions underlying the analysis of variance also apply to covariance. Treatment effects, for example, should be constant and additive, and the experimenter would do well to check the variances for homogeneity before proceeding with the analysis.

In addition, when designing experiments to which covariance is to be applied, *it is very important to make certain that the control variable used is not in any way affected by the treatment.* This would quite obviously destroy any effectiveness of the control variable in terms of the theory underlying analysis of covariance.

Further uses of covariance will be found in books listed in the bibliography.

PROBLEMS

11.1 Three diets, A, B, and C, were compared for their effects on blood cholesterol level in human females. Age was used as a control factor because of its apparent association with cholesterol level. Following a specified time, the blood cholesterol level in mg/100 ml was determined for each subject. Referring to the data presented in Table 11.15,

a) Ignore the control variable and test for statistical differences among the cholesterol levels (Y-values) of the three groups by analysis of variance. (See Section 8.6.)

TABLE 11.15

Diet A			Diet B			Diet C		
Subject number	X_A (Age)	Y_A (mg/ml)	Subject number	X_B (Age)	Y_B (mg/ml)	Subject number	X_C (Age)	Y_C (mg/ml)
1	40	190	1	41	201	1	41	201
2	47	205	2	30	187	2	32	192
3	28	178	3	58	226	3	57	215
4	51	215	4	48	222	4	49	202
5	50	202	5	57	220	5	36	197
6	48	208	6	26	184	6	30	195
7	27	181	7	50	208	7	34	194
8	47	201	8	44	199	8	42	200
9	42	192	9	39	196	9	61	248
10	38	193	10	29	185	10	40	198

b) Using the control variable, perform analysis of covariance on the data according to the method shown in Section 11.3.

c) Compare the results obtained in (a) with those obtained from (b).

11.2 Analyze the data in Table 11.16 for significant differences among Treatments 1, 2, and 3 (see Section 11.3).

TABLE 11.16

Treatment 1		Treatment 2		Treatment 3	
X_1	Y_1	X_2	Y_2	X_3	Y_3
80	96	61	40	73	39
64	76	59	79	44	22
57	63	60	61	48	32
51	76	53	47	51	57
57	88	56	69	43	28
		50	49	50	47
		45	40	41	13
				51	56
				62	22

11.3 An experiment was performed to compare two drugs, A and B, in terms of effectiveness in lowering systolic blood pressure. Twenty subjects were randomly assigned to two groups of ten each. The blood pressures in both groups were measured both prior to and following drug administration. Analyze the data in Table 11.17 for a statistically significant difference between drugs A and B. (See Section 11.3.)

TABLE 11.17

A		B	
X_A(Predrug)	Y_A(Postdrug)	X_B(Predrug)	Y_B(Postdrug)
160	151	150	127
145	138	146	133
149	139	150	132
170	168	180	165
150	147	170	154
175	170	165	150
190	182	172	162
160	162	155	150
180	170	160	142
172	161	165	147

11.4 An experiment was performed to compare the effects of two factors and to determine the degree of interaction. One factor consisted of Treatments A and B, and the other factor was sex. Measurements were taken before and after treatment and there were five replications per combination of factors. Analyze the data in Table 11.18 for significance involving (1) treatment, (2) sex, and (3) interaction between the treatment and sex (see Section 11.4).

TABLE 11.18

Treatment A				Treatment B			
Males		Females		Males		Females	
X_A	Y_A	X_A	Y_A	X_B	Y_B	X_B	Y_B
12	40	11	60	18	54	25	64
15	53	20	75	18	55	14	61
26	63	6	58	14	66	5	53
28	58	25	63	16	54	15	60
12	54	8	54	6	49	20	55

CHAPTER TWELVE

Nonparametric Tests

12.1 DISTRIBUTION-FREE METHODS

Up to now, we have been concerned with *parametric* tests of significance, where sample statistics have been used to infer parameters, or population values.

Parametric tests assume that the variable in question follows, or at least approximates, some kind of distribution (normal, binomial, Poisson, etc.), and this distribution is then a basis for sample-to-population inference. Therefore, underlying every parametric test is an assumption concerning the distribution of the population data.

Since it is not always possible to make this assumption, and indeed, in some examples of biological data it is *obvious* that it cannot be made, it is sometimes appropriate to analyze data with so-called *nonparametric*, or *distribution-free* methods.

Such tests do not require knowledge about the population distribution. Further, they are usually simple to perform, and require a minimum of calculations. However, nonparametric tests should not be used simply because they are quicker and easier than standard parametric tests. They should be used only where assumptions cannot be made relative to population distributions, or possibly as rapid tests on small amounts of data derived from small pilot studies. In the latter case, they should be considered as a prelude to more rigorous testing with standard parametric devices.

The investigator who indiscriminately substitutes a distribution-free method for a parametric test may be sacrificing power. Power is defined here as the ability of a test to detect false hypotheses, and parametric tests in general have a greater sensitivity to false hypotheses. In other words, the probability of making a Type II error is smaller when a standard parametric test is used.

The following are intended as general guidelines for making a choice between a parametric method and a nonparametric test:

1. If it is obvious that no assumption can be made concerning the type of distribution that a variable follows or approximates, a nonparametric test is indicated. The same is true if ordinal-scale data are involved.

2. Nonparametric tests may be useful for obtaining rapid estimates on small amounts of trial data.

3. When making inferences and broad generalizations from sample data, parametric tests are usually preferred.

12.2 THE SIGN TEST FOR MATCHED PAIRS

A good example of the "quick and dirty" methods described in this chapter is the *sign test*. This test may be used as a substitute for the matched-pair *t*-test described in Section 6.6, particularly when there is ample reason to doubt the normality of the data. The computations are based on the binomial distribution, and they are very simple.

The following example will illustrate the sign test, and the data are contrived to show its major pitfall.

Example. Twenty dogs are tested for lymphocyte count before and after administration of a drug that is supposed to produce a decrease in circulating lymphocytes. The resulting data are expressed in thousands per cubic millimeter of blood in Table 12.1.

1. Note that each replication is assigned a (+) or (−), depending on whether the lymphocyte count increased or decreased following administration of the drug.

2. The two cases that did not change were assigned zeros. These are removed from the total *n* number and are not included in the calculations. The working *n*-number is therefore 18.

3. If the drug has *no effect* on lymphocyte count, then the probabilities of obtaining a plus value and a minus value in any randomly selected case will be equal. Therefore, $P(+) = \frac{1}{2}$, and $P(-) = \frac{1}{2}$. We can ignore the probability of obtaining a zero, since zeros are removed from the computation.

4. We then have a binomial distribution of *minus* values. This is the distribution of interest, because we wish to know whether the drug significantly *lowers* the lymphocyte count.

TABLE 12.1

Animal	Before	After	Sign
1	2.5	2.6	+
2	1.2	1.5	+
3	2.9	2.9	0
4	3.1	2.0	−
5	3.1	2.3	−
6	1.1	1.3	+
7	1.5	1.6	+
8	4.1	3.1	−
9	2.1	1.4	−
10	2.4	2.5	+
11	1.3	1.4	+
12	2.8	2.8	0
13	3.5	2.4	−
14	3.6	2.1	−
15	1.1	1.3	+
16	1.6	1.7	+
17	4.2	3.2	−
18	2.2	1.5	−
19	2.5	2.1	−
20	1.3	1.1	−

5. The distribution mean is np, and the standard deviation is $\sqrt{npq}$. Substituting, we have

$$\text{Mean} = \tfrac{1}{2}(18) = 9,$$
$$\text{Standard deviation} = \sqrt{\tfrac{1}{2} \times \tfrac{1}{2}(18)} = \sqrt{4.5} = 2.12.$$

6. Recall that we obtained ten minus values. Now we need to compute the probability of obtaining *ten or more minus values by chance alone*. Taking 9.5 as the lower limit of 10, we obtain

$$Z = \frac{9.50 - 9.00}{2.12} = 0.235.$$

Consulting Table IV, we find that 0.235 is considerably less than is needed to reject the null hypothesis when a one-tailed test is assumed.

7. The probability of obtaining ten or more minus values by chance alone is quite high; we have therefore found no statistical evidence that the drug is effective in reducing the lymphocyte count.

At this point we might try a matched-pair t-test and compare results. If we do, we will obtain a t-value of 2.91, which is significant beyond the 0.01 level! In other words, the t-test produces exactly the opposite of the results obtained from the sign test in the preceding example. Why the contradiction?

Looking back at the original data, it may be seen that a $(+)$ or $(-)$ sign merely represents a *direction* of change and does not take into consideration the *magnitude* of the change. A casual inspection of the data used in the example reveals that upward changes in lymphocyte count were usually quite small, while downward changes were substantially greater. The sign test is not sensitive to these differences in magnitude but, since the computations of the t-test incorporate *absolute* changes, it produced significant results.

Thus it may be seen that, although the sign test has the advantage of simplicity, the careful investigator will do well to scan the data in order to determine whether absolute changes have reasonably similar magnitudes in both directions.

12.3 THE WILCOXON TEST FOR UNPAIRED DATA

The Wilcoxon test for unpaired data is a useful nonparametric substitute for the t-test considered in Section 6.7. As nonparametric tests go, the Wilcoxon test is reputed to have a high degree of efficiency.

This test is based on a ranking system, and the following simple example will serve to illustrate the basic computations.

Example. The pulse rates of six males were compared with the pulse rates of six females, and the following data were obtained:

Males	74	77	78	75	72	71
Females	80	83	73	84	82	79

1. As a first step, we rank *all* the above data in order of increasing magnitude. As we do, we underline those values taken from the group with the *smallest* mean. By simple inspection, this is quite obviously the male group. Thus we have:

$$\underline{71} \quad \underline{72} \quad 73 \quad \underline{74} \quad \underline{75} \quad \underline{77} \quad \underline{78} \quad 79 \quad 80 \quad 82 \quad 83 \quad 84$$
$$\underline{1} \quad \underline{2} \quad 3 \quad \underline{4} \quad \underline{5} \quad \underline{6} \quad \underline{7} \quad 8 \quad 9 \quad 10 \quad 11 \quad 12$$

2. Under each ordered value we also show a number indicating its rank, with the number 1 assigned to the lowest rank. Note that the rank numbers are underlined, or not, according to the raw data they represent.

3. Now it may be seen that, if there is actually *no difference* between the male and female groups, the values of the ranks should be such that the rank totals are *approximately equal.* In our case in point, we have

Males: $\underline{1} + \underline{2} + \underline{4} + \underline{5} + \underline{6} + \underline{7} = 25,$
Females: $3 + 8 + 9 + 10 + 11 + 12 = 53.$

In this situation the rank total of the female group is much higher than that of the males, indicating that *larger* rank numbers make up that total.

4. We now compute the value U by

$$U = T_1 - \tfrac{1}{2}n_1(n_1 + 1),$$

where T_1 is always the total of the ranks assigned to the group with the *smallest* mean, n_1 is the number of cases in the group with the smallest mean, and n_2 is the number of cases found in the group with the largest mean. Thus

$$U = T_1 - \tfrac{1}{2}n_1(n_1 + 1)$$

$$= 25 - \tfrac{1}{2}(6)(7)$$

$$= 25 - 21 = 4.$$

5. Turning to Table IX in the appendix, we look down the "n_1-n_2" column to the point indicated by $n_1 = 6$, $n_2 = 6$. The $C_{n_1 n_2}$ column at this point shows the figure 924, which represents the number of possible combinations of ranks from a total of 12, taken 6 at a time. We now find the U-value, 4, along the top portion of the table, and going down the column headed by 4 to the point opposite 924, we find the value 12. This represents the number of possible rank totals that would be equal to or less than 25. The probability, therefore, of finding a rank total of 25 or less in this case is

$$\tfrac{12}{924} = 0.012.$$

6. If we assume a one-tailed test, we may conclude significance beyond the 0.05 level. If a two-tailed test were involved, we would need to double the probability, obtaining 0.024, and we could still claim significance beyond the 0.05 level.

What happens when there are two or more values alike, producing ties? The following example illustrates the procedure used to handle ties and also shows that n_1 does not have to be exactly equal to n_2.

Example. We are given the following set of data:

n_1:   19   14   20   17   18   20
n_2:   24   19   33   32   21   18   20

1. We again arrange all units of data in the order of increasing value. Thus we have 13 pieces of data arranged as follows:

<u>14</u>   <u>17</u>   <u>18</u>   18   <u>19</u>   19   <u>20</u>   <u>20</u>   20   21   24   32   33

2. We again underline those values obtained from n_1, which is the set with the smallest mean. This time, however, the situation is complicated by a tie of three values (20) and two ties of two values each (18 and 19). We therefore assign to each member of a tie the *mean* value of the ranks that would normally be assigned if there were no tie! Thus we have

<u>14</u>	<u>17</u>	<u>18</u>	18	<u>19</u>	19	<u>20</u>	<u>20</u>	20	21	24	32	33
1	2	3.5	3.5	5.5	5.5	8	8	8	10	11	12	13

Note that the rank assigned to both 18's is the mean of 3 and 4, or 3.5. The rank assigned to both 19's is 5.5, which is the mean of 5 and 6. Each value of 20 is assigned the rank 8, which is the mean of 7, 8, and 9. Values that *follow* tied ranks are assigned the rank they would have received if there were no ties.

3. Adding the rank numbers of n_1 and n_2, we have

$$n_1: \quad 1 + 2 + 3.5 + 5.5 + 8 + 8 \qquad\qquad = 28,$$
$$n_2: \quad 3.5 + 5.5 + 8 + 10 + 11 + 12 + 13 = 63.$$

4. T_1, which is the total of the ranks obtained from the set with the *smallest* mean, is 28.

5.
$$U = T_1 - \tfrac{1}{2}n_1(n_1 + 1) = 28 - \tfrac{1}{2}(6)(7)$$
$$= 28 - 21 = 7.$$

6. Turning to Table IX, we find that when $n_1 = 6$ and $n_2 = 7$, $C_{n_1 n_2}$ is 1716. Looking down the U column headed by 7, we find 44 as the number of combinations of ranks that would produce a total of 28 or less. Our probability value is therefore

$$P = \tfrac{44}{1716} = 0.025$$

for a one-tailed test. If a two-tailed test is appropriate, the probability statement would be 0.05.

7. On the basis of either a one- or two-tailed test, we would reject the null hypothesis that no significant difference exists between the rank totals derived from n_1 and n_2.

12.4 RANK CORRELATION

In Chapter 10 we discussed the concept of correlation and the computation of the sample coefficient r, which you will recall is known as the Pearson product-moment r. This, of course, is a parametric test, which assumes that both the X and Y variables are *random* variables and are normally distributed in the population.

In cases where this assumption cannot be made, a nonparametric correlation called *rank* correlation (R) may be used. This method may also be used in situations where one of the variables may itself be derived from a ranking system. For example, flower color might be expressed as 1, 2, 3, ..., as ranks that are assigned to shades ranging from light to dark.

Like most nonparametric techniques, the computation of a rank correlation is simpler than the calculation of its parametric counterpart. The following example will illustrate the procedure.

TABLE 12.2

| Subject | Hb (mg/100 ml) | | RBC (millions/mm³) | | | |
	X	Rank	Y	Rank	$(X - Y)$	$(X - Y)^2$
1	15.2	(5.5)	5.1	(4)	±1.5	2.25
2	16.4	(1)	5.4	(2)	−1.0	1.00
3	14.2	(11)	4.5	(9)	+2.0	4.00
4	13.0	(12)	4.2	(12)	0.0	0.00
5	14.5	(10)	4.3	(10.5)	−0.5	0.75
6	16.1	(2)	6.1	(1)	+1.0	1.00
7	15.2	(5.5)	5.2	(3)	+2.5	6.25
8	14.8	(8)	4.3	(10.5)	−2.5	6.25
9	15.8	(3)	4.7	(7)	−4.0	16.00
10	14.9	(7)	4.8	(5.5)	+1.5	2.25
11	15.6	(4)	4.6	(8)	−4.0	16.00
12	14.7	(9)	4.8	(5.5)	+3.5	12.25
$n = 12$					$(X - Y)^2 = 68$	

Example. Suppose that we wish to determine whether an association exists between hemoglobin content in mg/100 ml and red-blood-cell count in millions per cubic millimeter. We randomly draw 12 adult males from the population and measure hemoglobin concentration and red-cell count on each individual (our results appear in Table 12.2). As we emphasized in Chapter 10, both variables are *random* but, for convenience, we will use the symbol (X) for hemoglobin concentration and (Y) for red cell count. Since we suspect that our data do not reflect a normally distributed population, we will use the rank-correlation procedure as follows:

1. Looking at Table 12.2, you will note that we have ranked each measurement within each of the X and Y columns, assigning 1 to the highest number, 2 to the next highest, and so on. Where ties exist, as in

the case of the two 15.2 values in the X column, we "split the difference." For example, ordinarily the first 15.2 would be ranked 5 and the second 15.2 would be ranked 6. The mean of 5 and 6 is 5.5 so we assign this rank to *both* values. The next lowest value of X is then ranked 7, and so on.

2. Since there are relatively few ties in our data, R can be calculated quickly as

$$R = 1 - \frac{6\Sigma(X - Y)^2}{n(n^2 - 1)} = 1 - \frac{6(68)}{12(144 - 1)} = 0.76.$$

3. There could be a significant number of ties among the data, especially if one of the variables itself consists of ranks (as in the shades of flower color, quality of mouse nests, etc.) If so, the computations are longer, but certainly not difficult. The first step is to compute the correction factor for ties by using the formula

$$\text{Ties} = \Sigma \frac{m^3 - m}{12},$$

where 12 is a constant and does *not* refer to the n-number in our particular example. The letter m represents the number of measurements with the same value found in the X column or in the Y column. For example, in the X column in Table 12.2, there are two measurements with the same value (15.2). In the Y column there are two measurements with the value 4.8, and two with the value 4.3. Calculating the correction factor for ties for *each column*, we have

$$\text{Ties}(X) = \frac{(2)^3 - 2}{12} = 0.50$$

and

$$\text{Ties}(Y) = \frac{2^3 - 2}{12} + \frac{2^3 - 2}{12} = 1.00.$$

We then calculate for X,

$$X = \frac{n^3 - n}{12} - \text{Ties}(X)$$

$$= \frac{(12)^3 - 12}{12} - 0.50 = 142.5,$$

and for Y,

$$Y = \frac{n^3 - n}{12} - \text{Ties}(Y)$$

$$= \frac{(12)^3 - 12}{12} - 1.00 = 142.$$

Finally, we can calculate R as

$$R = \frac{X + Y - \sum(X - Y)^2}{2\sqrt{(X)(Y)}}$$

$$= \frac{142.5 + 142 - 68}{2\sqrt{(142)(142.5)}} = 0.76.$$

As you can see, there is no appreciable difference between the rank-correlation coefficient calculated the long way and the one calculated by the simpler formula that assumes few or no ties. Given data with a significant number of ties, however, we would see a meaningful difference.

4. In Chapter 10, we stressed that a sample correlation coefficient could represent a chance deviation from zero; we therefore emphasized the necessity of testing it for "significance." When using the rank-correlation procedure, R can be tested for significance by the following formula, provided the n-number is at least 10:

$$t = R\frac{n - 2}{\sqrt{1 - R^2}}$$

$$= 0.76 \times \frac{10}{\sqrt{0.4224}} = 11.69,$$

which, at $(n - 2)$, or 10 degrees of freedom, is statistically significant beyond the 0.001 level.

12.5 RUNS

The technique of *runs* has useful applications to biological problems. An ecologist may wish to know whether annual fluctuations in a population are random from one year to another, or whether they appear to follow a pattern, possibly because of environmental influences. Or it might be desirable to determine whether the number of cases of a certain disease fluctuates randomly from one year to another, or whether nonrandom changes in incidence might be found, and thus associated with conditions apparently conducive to increases or decreases in the number of cases.

The technique is extremely simple, as illustrated by the following example.

Example. The numbers of cases of malaria reported from a certain area over a period of successive years were 51, 82, 64, 32, 11, 12, 54, 71, 90, 101, 84, 72, 45, 20, 74, 15. Did the number of cases fluctuate in a random manner or is there some kind of pattern?

1. First, we establish a reference point from which to work. Usually, the most convenient reference point is the median, which, in this case, is 64.

2. Now we simply assign the letter "*a*" to those values under 64 and "*b*" to those above 64. Any value at the median will be discarded and will not be included in the data.

3. Substituting the appropriate letter for each number in the original distribution, we have

$$\underline{a}, \quad \underline{b}, \quad \underline{aaaa}, \quad \underline{bbbbb}, \quad \underline{aa}, \quad \underline{b}, \quad \underline{a}.$$

Note that each run of $\underline{a}$ or $\underline{b}$ is underlined and that a single $\underline{a}$ or $\underline{b}$ is considered a *run*.

4. We now need three values: (1) the number of $\underline{a}$'s, (2) the number of $\underline{b}$'s and (3) the total number of *runs*. Thus

$$\underline{a} = 8, \qquad \underline{b} = 7, \qquad \text{runs} = 7.$$

5. Consulting Table X in the appendix, we go across the top row for $a = 8$, and down the left margin for $b = 7$. At the point of intersection of the row and column, we find the numbers 4 and 13. This tells us that if the number of runs is 4 *or less* we can assume that random fluctuation did *not* occur. In other words, the *a*'s and *b*'s are ordered in such a way as to cause us to suspect that some external factor has produced that orderliness. The same would be true if the number of runs found is 13 *or more*. This would also indicate an orderly alternation of *a*'s and *b*'s and would be inconsistent with randomness.

6. Since our example yielded seven runs, this number is neither low enough nor high enough to permit a conclusion of nonrandomness. We must therefore assume that the incidence of malaria fluctuated from one year to another in a random manner.

Another application of "runs" is illustrated by the following.

Example. An ecologist counted the number of birds in a specified area by keeping careful track of territories. In successive years he made the observations shown below. Does the size of the population fluctuate randomly from one year to another, or it there a self-limiting mechanism which produces a definite pattern?

Year	1954	1955	1956	1957	1958	1959	1960	1961	1962	1963
Number	13	13	12	10	8	11	9	7	6	5

1. First we determine the median value to be 9.5. Assigning the letter "*a*" to values lower than 9.5 and the letter "*b*" to values larger than 9.5 produces

$$\underline{aaaa} \quad \underline{b} \quad \underline{a} \quad \underline{bbbb}$$

$$a = 5$$

$$b = 5$$

$$\text{runs} = 4$$

2. Consulting Table X, we find the numbers 2 and 10 at the intersection of $a = 5$, $b = 5$. Since the actual number of runs obtained is 4, this value is neither small enough nor large enough for us to conclude that any kind of pattern other than random fluctuation exists in the annual shifts in the bird population.

12.6 THE KRUSKAL–WALLIS TEST

In Chapter 8 we described the completely randomized design as a way to compare three or more groups. The completely randomized design was analyzed by analysis of variance, which is a parametric test that is "tied to" a population distribution and therefore assumes certain properties such as normality and homogeneity of treatment variances.

If the assumptions of analysis of variance cannot be met, the *Kruskal–Wallis* nonparametric test may be used instead. The Kruskal–Wallis is especially useful for analyzing data that involve more than two treatments. Also, in Chapter 1 we briefly discussed an experiment in which the investigator rated the quality of mouse nests by using an arbitrary scale such as 1, 2, 3, and 4. The use of this *ordinal* scale is helpful in situations where the experimenter wants to express qualitative responses in quantitative terms. Data expressed on the ordinal scale cannot, however, be analyzed by parametric methods. The Kruskal–Wallis test would probably be the appropriate statistical test in situations that involve the ordinal scale.

As with most nonparametric tests, the computations associated with the Kruskal–Wallis test are relatively simple, as illustrated by the following example:

Example. As part of a general study of phytoplankton population dynamics in a small lake, a researcher compared the percent cell number of *Rhodomonas minuta* found at a specific site during spring, midsummer, and late fall. Ten replicate samples were obtained during each of the three seasonal collections, yielding the data shown in Table 12.3. Note that the raw data are arranged in an array, from lowest to highest.

TABLE 12.3

Spring		Midsummer		Late Fall	
%	Rank	%	Rank	%	Rank
9.6	10.5	4.8	1	5.4	2
11.2	14	7.6	5.5	6.5	3
11.6	16	7.6	5.5	7.1	4
11.7	17.5	9.2	9	8.0	7
12.8	19	9.6	10.5	8.8	8
12.9	20	21.1	22	9.5	9
15.8	21	24.6	24.5	10.2	12
22.7	23	25.6	26	10.7	13
24.6	24.5	26.4	27	11.3	15
32.5	28	32.8	29	11.7	17.5
$n_S = 10$ $R_S = 193.5$		$n_M = 10$ $R_M = 353.5$		$n_F = 10$ $R_F = 90.5$	

1. The first step in our computations is to rank the thirty (30) measurements that comprise the data, beginning with the smallest measurement, which is 4.8. If ties should occur, either within or between groups, the measurements that make up the tie are assigned the mean of the ranks that would be assigned to each one if there were no tie. For example, look at the "midsummer" column and note that 7.6 would be ranked 5 and the next higher measurement would be ranked 6. However, since there are two measurements with a value of 7.6, they are *both* assigned the mean of ranks 5 and 6, or 5.5. Now that we have "used" ranks 5 and 6, the next higher measurement (8.0) is ranked 7, and so on.

2. The next step is to sum the ranks assigned to each treatment, yielding R_S, R_M, and R_F. We should also note that n_T is 30 and that n_S, n_M, and n_F are each 10. We also have three treatment groups, so $k = 3$.

3. We are now ready to substitute our data in the Kruskal–Wallis formula as follows:

$$H = \frac{12}{n_T(n_T + 1)} \left[\frac{R_S}{n_S} + \frac{R_M}{n_M} + \frac{R_F}{n_F} \right] - 3(n_T + 1)$$

$$= \frac{12}{30(31)} \left[\frac{(193.5)^2}{10} + \frac{(353.5)^2}{10} + \frac{(90.5)^2}{10} \right] - 3(31)$$

$$= 128.77.$$

4. Since H is distributed as chi-square, we enter Table V with $(k - 1)$, or 2 degrees of freedom. The value of H is significant well beyond the 0.005 level, and we therefore conclude that the evidence strongly suggests that there are significant differences among the seasonal percent number of *Rhodomonas minuta*.

5. In some situations, the data to be analyzed by the Kruskal–Wallis test may contain a large number of ties. This may be especially true if ordinal data are involved. In such cases, a correction factor should be applied. In our example it would be done as follows:

   a) Let m stand for the number of measurements that have the same specific value. For example, in our data we had two numbers with a value of 9.6, two with a value of 11.7, two with a value of 7.6, and two with a value of 24.6.

   b) Calculate $(m^3 - m)$ for each tie and then obtain $\sum(m^3 - m)$. We therefore have

   $$\text{Ties} = (8-2) + (8-2) + (8-2) + (8-2) = 24.$$

   c) Then the correction factor is:

   $$(\text{CF}) = 1 - \frac{\text{Ties}}{n_T^3 - n_T}$$

   $$= 1 - \frac{24}{27000 - 30} = 0.999.$$

   d) We now compute the corrected H-value by

   $$\frac{H}{\text{CF}} = \frac{128.77}{0.999} = 128.90,$$

   which obviously makes very little difference as far as our example is concerned because we had very few ties among the data.

An important point to remember relative to the Kruskal–Wallis test is the fact that the test rapidly loses power as the treatment n-numbers decrease down to five or below. Also, in this situation, H is no longer distributed like chi square, and it must be looked up in special tables. A better solution, obviously, is to *plan ahead* and ensure that an adequate number of replications is obtained for each treatment.

12.7 THE FRIEDMAN χ_r^2 TEST

In Chapter 9 we described the randomized-block design. Like other designs that are analyzed by analysis of variance, the randomized-block design requires that the assumptions of analysis of variance be met to at least a reasonable degree. If it cannot be assumed that the data are normally distributed in the population, or if ordinal data are involved, then a nonparametric version of the randomized-block design may be used. This test, called the Friedman χ_r^2 test, is illustrated in Table 12.4, using the data found in Chapter 9, page 151.

TABLE 12.4

Test	\multicolumn{7}{c	}{Blocks (litters)}	
	1 Rank	2 Rank	3 Rank	4 Rank	5 Rank	6 Rank	7 Rank	
P	5.4 (2)	4.0 (2)	7.0 (3)	5.8 (2)	3.5 (1)	7.6 (2)	5.5 (2)	$\sum R_P = 14$
A	6.0 (3)	4.8 (3)	6.9 (2)	6.4 (3)	5.5 (3)	9.0 (3)	6.8 (3)	$\sum R_A = 20$
B	5.2 (1)	3.9 (1)	6.5 (1)	5.6 (1)	3.9 (2)	7.0 (1)	5.4 (1)	$\sum R_B = 8$

$k = 3$ $\qquad\qquad\qquad\qquad\qquad\qquad\qquad\qquad\qquad\qquad n = 7$

1. We begin by ranking the treatment measurements within each block, and then obtain the sum of the ranks for each treatment, yielding $\sum R_P$, $\sum R_A$, and $\sum R_B$. Since we have three treatments, $k = 3$, and $n = 7$ because there are seven replications per treatment (or seven blocks).

2. We now substitute our data into the Friedman χ_r^2 formula, as follows:

$$\chi_r^2 = \frac{12}{nk(k+1)}\left[\left(\sum R_P\right)^2 + \left(\sum R_A\right)^2 + \left(\sum R_B\right)^2\right] - 3n(k+1)$$

$$= \frac{12}{7(3)(4)}[(14)^2 + (20)^2 + (8)^2] - 3(7)(4)$$

$$= 10.38.$$

3. The Friedman χ_r^2 statistic is distributed as χ^2 with $(k-1)$ degrees of freedom. We therefore enter Table V at two degrees of freedom and find that our obtained value of 10.38 is statistically significant beyond the 0.01 level. If you refer back to Chapter 9, page 153, you will find that we also rejected the null hypothesis when we analyzed the same data with analysis of variance.

12.8 SUMMARY

Nonparametric statistics is a rapidly growing field, and theoretical statisticians are bending their efforts toward designing new and better nonparametric tests that will have the power of parametric procedures. We occasionally hear predictions that such standard statistical procedures as Student's t and analysis of variance will some day be replaced by non-parametric tests. As of now, however, parametric tests are still the most widely used and apparently still the most dependable procedures.

As you come (thankfully, no doubt) to the end of this book, you should have gained a basic knowledge of experimental statistics and design that will allow you to use these procedures wisely. In addition, by now you should be able to intelligently consult more advanced textbooks and references as your needs dictate since, quite obviously, in a book of this length much has been left unsaid. Perhaps of greatest importance is the valuable exposure that you have had to the orderly methods of reasoning upon which statistical thought and procedures are based.

PROBLEMS

12.1 Use the sign test to analyze the set of matched-pair data shown in Table 12.5, for significant increase in the values derived from treatment B over those produced by treatment A (see Section 12.2).

TABLE 12.5

Subject	Treatment A	Treatment B
1	46	52
2	41	43
3	37	37
4	32	32
5	28	31
6	43	39
7	42	44
8	51	53
9	28	26
10	27	31

12.2 Two laboratory methods of determining the level of protein-bound iodine are compared by using 12 female adults. Use the sign test to determine whether the results (shown in Table 12.6) associated with method B are significantly higher than the values produced by method A (see Section 12.2).

12.3 Two ponds were compared in terms of alkalinity, expressed in milligrams per liter. Five samples were taken from each pond. Use the Wilcoxon two-sample rank test to determine whether a significant difference in alkalinity exists between the ponds (see Section 12.3).

Pond I: 102, 116, 122, 112, 104
Pond II: 108, 117, 115, 120, 105

TABLE 12.6

Subject	Method A	Method B	Subject	Method A	Method B
1	4	5	7	9	10
2	6	5	8	3	4
3	3	3	9	4	5
4	5	6	10	4	5
5	7	8	11	5	5
6	8	6	12	4	6

12.4 Use the Wilcoxon two-sample rank test to determine whether a significant difference exists between the following two sets of data (see Section 12.3).

A: 82, 86, 30, 21, 38, 29, 29, 19
B: 124, 116, 54, 54, 110, 29, 39, 54

12.5 Apply the rank correlation technique to the paired data shown in Table 12.7 (see Section 12.4).

TABLE 12.7

	X	Y		X	Y
1	21	25	6	63	64
2	21	20	7	84	81
3	42	47	8	84	83
4	42	44	9	105	106
5	63	61	10	105	111

12.6 Apply the rank-correlation technique to the paired data (see Section 12.4) shown in Table 12.8.

TABLE 12.8

	X	Y		X	Y
1	24	105	6	33	140
2	36	213	7	38	173
3	48	274	8	27	136
4	34	198	9	31	155
5	28	128			

12.7 The following data represent annual counts of shore birds observed at a specific location on a lake shore during the migratory season. Use the technique of "runs" to determine whether the fluctuation from one year to another is random, or whether a pattern is present (see Section 12.5).

Year	1956	1957	1958	1959	1960	1961	1962	1963	1964	1965	1966
Number	489	795	540	680	970	320	830	293	565	460	602

12.8 The following data represent annual catches of fish at a specific location in one of the Great Lakes; the data are expressed in tons. Analyze these data to determine whether the annual fluctuation is random or whether a pattern is evident (see Section 12.5).

Year	1943	1944	1945	1946	1947	1948	1949	1950	1951	1952	1953
Number	56	57	61	59	62	52	56	53	51	49	48

Year	1954	1955	1956	1957	1958	1959	1960	1961	1962	1963	1964
Number	52	50	48	52	51	42	49	53	40	43	42

12.9 An experiment was performed to determine whether significant differences exist among the mean critical thermal maximum (CTM) values of fish reared at different temperatures (7, 15, and 25°C). The data in Table 12.9 were obtained from ten replications per rearing temperature. Use the Kruskal–Wallis test to determine whether significant differences are present among the three groups. (See Section 12.6.) (Data adapted from an experiment performed by J. Spotila and K. Hassan.)

TABLE 12.9

Rearing temperature		
25°C	*15°C*	*7°C*
36.1	33.2	29.3
35.6	33.0	27.9
35.5	34.4	28.5
34.8	32.5	34.0
34.7	30.9	29.5
34.6	33.6	28.9
33.8	30.9	29.2
33.5	30.7	28.1
33.2	32.9	27.8
34.6	29.8	27.2

12.10 Three methods of determining prothrombin levels were tested, using single donors as blocks. Analyze the data in Table 12.10 with the Friedman χ_r^2 test for differences among methods A, B, and C. Prothrombin levels are expressed in units/ml. (See Section 12.7.)

TABLE 12.10

Methods	Donors (Blocks)							
	1	*2*	*3*	*4*	*5*	*6*	*7*	*8*
A	49	84	75	91	84	81	49	58
B	52	86	85	95	91	86	56	60
C	50	75	72	92	87	81	52	68

Bibliography

Bibliography

The following books are suggested for additional reading and reference. The list is divided into two parts: (1) those recommended for collateral reading with this text, and (2) those recommended for further exploration of statistical procedures and for use as references.

RECOMMENDED FOR COLLATERAL READING

Bahn, Anita, K., *Basic Medical Statistics*. New York: Grune and Stratton, 1972. This book is of special interest to those who are primarily concerned with applications of statistics to medicine. It is in paperback form, and contains numerous worked-out examples of problems likely to be encountered in the clinical or laboratory situation.

Hartl, Daniel L., *Our Uncertain Heritage*. New York: J. B. Lippincott, 1977. An understandable text in human genetics that is aimed at the nonscience student. Of special interest to the reader who would like to further explore the concepts discussed in Chapter 4 of this text.

Huff, D., *How to Lie With Statistics*. New York: W. W. Norton, 1954. A delightful little book that describes some of the less creditable ways in which statistics is used—and misused. Written in an informal and entertaining style, it is a fascinating exposé of statistical charlatanism. Available in most public libraries, it can easily be read in an evening.

Stine, Gerald J., *Biosocial Genetics*. New York: Macmillan, 1977. An excellent book on human genetics that ties principles of genetics to social issues. Includes problems involving genetic counseling and genetic screening. A good choice for reading after the material in Chapter 4 of this text has been digested.

Weaver, Warren, *Lady Luck*. New York: Doubleday, 1963. A very readable paperback for those who hate mathematics but would like to learn how the student, the scientist, the housewife, and the gambler can use basic probability theory.

Wilson, E. O., and Bossert, W. H., *A Primer of Population Biology.* Stamford, Conn.: Sinauer Associates, 1971. A small paperback filled with understandable information on quantitative aspects of ecology and genetics. Contains an easy introduction to the description of biological situations by the use of mathematical models.

RECOMMENDED FOR FURTHER READING AND FOR REFERENCE

Batschelet, E., *Introduction to Mathematics for the Life Sciences.* New York: Springer-Verlag, 1971. Useful to the biologist who wants a better mathematics background in those concepts that are pertinent to the field. Mathematical concepts related to a number of research areas are presented in an intuitive manner.

Remington, R. D., and Schork, M. A., *Statistics with Applications to the Biological and Health Sciences.* Englewood Cliffs, N.J.: Prentice-Hall, 1970. Especially useful as a reference for students of medicine, public health, nursing, or pharmacy. Includes material on demography and vital statistics.

Snedecor, G. W., *Statistical Methods.* Ames, Iowa: Iowa State University Press, 1956. A classic text written by a leading authority, the emphasis is on agricultural research problems. Not easy reading but an excellent reference.

Sokal, R., and Rohlf, F. J., *Biometry.* San Francisco: W. H. Freeman, 1969. An excellent reference text that should be on the bookshelf of everyone engaged in biological research. Presents a comprehensive coverage of general biometrics with numerous worked-out examples. Accompanied by a separate booklet of statistical tables. Highly recommended for those already involved or potentially involved in biological research.

Steel, R., and Torrie, J. H., *Principles and Procedures of Statistics.* New York: McGraw-Hill, 1960. Another excellent reference that has become a classic. Contains considerable information on transformations and multiple-range tests.

Wadley, F. M., *Experimental Statistics in Entomology.* Washington, D.C.: Graduate School Press, U.S. Department of Agriculture, 1967. Presents experimental designs and statistical methods of special interest to the entomologist. The emphasis is on practical applications.

Woolf, Charles M., *Principles of Biometry.* Princeton: D. Van Nostrand, 1968. An understandable text that also serves as a good reference. Contains a chapter on bioassay and a good explanation of Model II ANOVA.

Zar, Jerrold H., *Biostatistical Analysis.* Englewood Cliffs, N.J.: Prentice-Hall, 1974. A comprehensive textbook that covers the standard concepts and tests. In addition, it treats concepts such as the diversity index and stepwise multiple regression. Contains a large section of statistical tables. Strongly recommended for further reading as well as for a permanent reference.

Appendix

TABLE I Powers and Roots

No.	Sq.	Sq. Root	Cube	Cube Root	No.	Sq.	Sq. Root	Cube	Cube Root
1	1	1.000	1	1.000	51	2,601	7.141	132,651	3.708
2	4	1.414	8	1.260	52	2,704	7.211	140,608	3.733
3	9	1.732	27	1.442	53	2,809	7.280	148,877	3.756
4	16	2.000	64	1.587	54	2,916	7.348	157,464	3.780
5	25	2.236	125	1.710	55	3,025	7.416	166,375	3.803
6	36	2.449	216	1.817	56	3,136	7.483	175,616	3.826
7	49	2.646	343	1.913	57	3,249	7.550	185,193	3.849
8	64	2.828	512	2.000	58	3,364	7.616	195,112	3.871
9	81	3.000	729	2.080	59	3,481	7.681	205,379	3.893
10	100	3.162	1,000	2.154	60	3,600	7.746	216,000	3.915
11	121	3.317	1,331	2.224	61	3,721	7.810	226,981	3.936
12	144	3.464	1,728	2.289	62	3,844	7.874	238,328	3.958
13	169	3.606	2,197	2.351	63	3,969	7.937	250,047	3.979
14	196	3.742	2,744	2.410	64	4,096	8.000	262,144	4.000
15	225	3.873	3,375	2.466	65	4,225	8.062	274,625	4.021
16	256	4.000	4,096	2.520	66	4,356	8.124	287,496	4.041
17	289	4.123	4,913	2.571	67	4,489	8.185	300,763	4.062
18	324	4.243	5,832	2.621	68	4,624	8.246	314,432	4.082
19	361	4.359	6,859	2.668	69	4,761	8.307	328,509	4.102
20	400	4.472	8,000	2.714	70	4,900	8.367	343,000	4.121
21	441	4.583	9,261	2.759	71	5,041	8.426	357,911	4.141
22	484	4.690	10,648	2.802	72	5,184	8.485	373,248	4.160
23	529	4.796	12,167	2.844	73	5,329	8.544	389,017	4.179
24	576	4.899	13,824	2.884	74	5,476	8.602	405,224	4.198
25	625	5.000	15,625	2.924	75	5,625	8.660	421,875	4.217
26	676	5.099	17,576	2.962	76	5,776	8.718	438,976	4.236
27	729	5.196	19,683	3.000	77	5,929	8.775	456,533	4.254
28	784	5.292	21,952	3.037	78	6,084	8.832	474,552	4.273
29	841	5.385	24,389	3.072	79	6,241	8.888	493,039	4.291
30	900	5.477	27,000	3.107	80	6,400	8.944	512,000	4.309
31	961	5.568	29,791	3.141	81	6,561	9.000	531,441	4.327
32	1,024	5.657	32,768	3.175	82	6,724	9.055	551,368	4.344
33	1,089	5.745	35,937	3.208	83	6,889	9.110	571,787	4.362
34	1,156	5.831	39,304	3.240	84	7,056	9.165	592,704	4.380
35	1,225	5.916	42,875	3.271	85	7,225	9.220	614,125	4.397
36	1,296	6.000	46,656	3.302	86	7,396	9.274	636,056	4.414
37	1,369	6.083	50,653	3.332	87	7,569	9.327	658,503	4.431
38	1,444	6.164	54,872	3.362	88	7,744	9.381	681,472	4.448
39	1,521	6.245	59,319	3.391	89	7,921	9.434	704,969	4.465
40	1,600	6.325	64,000	3.420	90	8,100	9.487	729,000	4.481
41	1,681	6.403	68,921	3.448	91	8,281	9.539	753,571	4.498
42	1,764	6.481	74,088	3.476	92	8,464	9.592	778,688	4.514
43	1,849	6.557	79,507	3.503	93	8,649	9.644	804,357	4.531
44	1,936	6.633	85,184	3.530	94	8,836	9.695	830,584	4.547
45	2,025	6.708	91,125	3.557	95	9,025	9.747	857,375	4.563
46	2,116	6.782	97,336	3.583	96	9,216	9.798	884,736	4.579
47	2,209	6.856	103,823	3.609	97	9,409	9.849	912,673	4.595
48	2,304	6.928	110,592	3.634	98	9,604	9.899	941,192	4.610
49	2,401	7.000	117,649	3.659	99	9,801	9.950	970,299	4.626
50	2,500	7.071	125,000	3.684	100	10,000	10.000	1,000,000	4.642

TABLE II Random Numbers

09 90 73 75 54	18 04 18 76 01	82 58 95 87 64	00 54 02 64 40
08 28 53 91 89	35 30 84 75 41	86 60 08 75 59	99 32 62 37 26
77 19 21 51 99	51 50 81 47 55	30 23 85 46 96	38 71 93 64 83
33 85 84 56 65	71 27 13 73 13	34 48 38 21 85	80 68 96 62 68
38 37 97 21 73	00 40 12 82 13	10 29 54 64 54	21 63 03 11 27
97 97 07 90 56	32 51 47 20 66	82 98 67 97 28	53 15 72 18 13
10 64 33 41 94	78 81 81 61 00	54 33 59 61 39	24 31 41 36 75
20 74 13 99 31	86 69 93 68 62	83 97 27 10 53	42 92 88 72 52
07 93 40 34 06	93 11 44 17 87	58 30 03 39 64	47 32 55 59 88
76 97 48 34 42	81 01 87 47 95	71 30 08 14 71	91 47 33 33 90
95 06 62 17 10	27 54 69 49 03	04 94 62 90 00	22 93 29 42 68
27 05 36 22 83	85 91 28 69 91	13 40 51 50 12	66 51 21 26 60
99 02 17 36 41	01 88 57 21 70	68 55 05 94 69	41 21 68 04 77
28 92 21 61 52	69 70 66 31 61	51 28 73 10 34	92 83 85 12 31
17 75 14 40 90	11 86 17 72 35	71 56 21 64 85	60 27 81 63 79
08 88 06 91 22	63 19 44 22 73	04 97 64 72 98	83 91 27 95 20
15 00 59 39 71	88 78 02 02 76	73 93 90 12 46	04 32 28 55 48
48 02 67 99 71	27 97 31 13 28	74 73 56 45 30	51 74 07 42 74
66 43 27 39 36	47 41 00 16 58	21 37 91 22 61	58 73 61 85 45
29 11 53 88 47	29 00 92 59 42	37 86 50 02 96	74 47 75 49 08
38 87 12 37 71	98 07 46 50 45	73 61 70 58 32	74 50 18 71 95
61 60 46 78 94	89 46 66 17 97	75 04 87 65 23	53 75 95 14 06
90 28 08 83 81	81 77 28 60 97	39 52 50 91 81	80 51 46 91 42
30 51 22 49 87	32 59 26 65 52	98 04 05 75 10	22 00 61 60 69
21 32 23 13 78	96 46 76 90 98	40 26 20 64 81	49 05 47 41 56

```
66 71 65 72 90   74 01 81 17 91   85 38 81 64 47   99 64 11 85 78   71 02 11 21 75
20 36 48 98 68   51 18 98 18 98   03 62 75 94 11   63 74 37 31 26   92 60 54 19 87
89 64 11 42 36   44 42 75 71 96   45 68 19 18 69   76 62 86 43 78   77 02 24 34 30
33 03 84 63 92   36 31 89 44 39   53 31 42 00 14   39 61 27 29 15   19 98 50 46 73
08 28 68 03 21   20 78 71 13 58   89 75 20 34 62   74 63 85 01 55   86 67 73 43 21

85 73 20 68 72   01 09 32 61 71   67 19 31 94 62   65 46 62 00 57   26 45 45 88 09
08 18 34 44 21   11 48 80 20 96   70 97 06 15 85   33 87 22 65 08   07 63 61 70 21
38 88 62 17 01   54 86 56 99 12   95 38 76 07 96   33 63 10 05 60   06 94 97 43 35
44 19 27 58 80   78 45 40 20 56   88 61 66 70 47   17 71 58 47 01   03 25 95 87 88
92 52 75 17 13   21 47 40 71 85   13 86 92 12 01   34 50 73 46 60   34 50 73 46 60

20 89 97 52 71   54 35 42 60 23   97 38 71 65 11   37 49 56 55 18   52 54 45 40 10
41 80 29 30 88   21 93 56 99 15   02 25 92 33 91   85 29 00 35 32   16 47 65 58 97
05 11 28 41 74   74 26 79 89 11   80 36 66 39 99   24 34 95 52 21   17 34 42 22 68
22 62 83 40 10   47 63 44 28 78   66 32 96 64 57   18 62 81 05 13   32 67 93 13 48
43 16 97 75 46   24 19 45 38 31   96 38 72 97 28   36 61 11 65 06   02 43 74 24 51

60 42 72 80 38   80 44 17 10 28   56 18 73 29 71   26 95 30 50 11   80 64 88 69 24
65 79 09 17 31   40 89 16 64 84   57 04 17 56 27   18 04 32 10 70   73 17 17 06 61
44 44 85 80 94   07 03 44 05 46   55 32 17 41 51   77 94 04 35 13   18 04 32 10 70
93 48 04 94 32   32 89 80 46 03   77 10 06 25 47   16 77 13 29 26   16 77 13 29 26
20 01 69 51 93   74 35 13 32 36   70 70 61 08 12   71 27 51 47 12   71 27 51 47 12

25 49 38 64 49   06 06 39 66 58   60 60 84 72 89   80 92 47 21 15   19 77 43 95 12
06 06 39 66 58
60 60 84 72 89
80 92 47 21 15
19 77 43 95 12
```

(continued)

TABLE II (continued)

94 74 62 11 17	66 54 30 69 08	27 13 80 10 54	60 49 78 66 44	41 94 41 50 41
01 10 88 74 94	06 24 94 17 34	76 02 21 87 12	31 73 62 69 07	46 55 61 27 39
54 88 08 81 40	74 49 55 03 58	74 51 73 56 75	14 97 65 21 12	88 93 57 39 68
68 82 78 21 56	27 10 75 74 89	35 43 62 20 73	28 14 15 39 80	51 75 03 31 05
74 22 73 02 00	92 30 89 03 75	84 38 92 04 26	24 84 94 86 91	49 59 60 13 04
32 88 95 80 60	95 45 31 86 35	85 54 98 90 26	37 92 16 99 07	49 49 64 41 90
44 57 16 58 47	04 54 73 99 84	30 06 52 39 62	30 00 45 83 36	55 67 11 79 67
44 07 05 04 80	35 77 25 59 18	18 61 52 16 91	14 39 39 70 29	41 85 45 48 00
82 40 92 18 33	26 08 72 03 57	89 52 43 11 90	26 80 46 05 77	79 31 86 68 82
77 15 21 67 43	80 18 60 07 71	77 43 35 05 87	78 86 14 82 03	94 19 60 61 89
59 25 22 17 25	46 59 47 94 08	29 47 24 57 24	45 76 39 81 76	14 70 90 24 40
82 70 30 71 85	78 84 67 30 10	49 72 43 41 47	99 66 01 23 44	92 31 85 78 90
09 49 49 05 25	05 99 00 47 55	06 46 22 10 28	04 87 49 24 74	43 20 06 18 20
61 10 03 96 89	64 61 76 18 99	97 67 48 63 87	32 32 70 49 25	96 56 46 96 50
63 35 14 21 05	87 69 54 03 87	14 33 96 68 79	42 09 66 87 37	50 82 18 83 69
64 01 72 06 57	09 61 46 26 87	73 47 43 53 30	17 59 83 09 98	95 66 80 55 95
65 75 87 55 21	97 45 37 82 11	03 43 27 85 54	37 20 01 50 52	29 98 62 41 08
42 51 71 40 63	15 92 62 50 22	54 14 75 63 02	45 21 20 49 49	40 63 05 18 30
58 47 73 78 96	94 16 53 55 14	12 39 88 07 78	20 19 98 64 78	05 40 17 56 67
43 50 34 50 18	81 47 66 11 76	07 05 74 43 86	03 73 32 12 31	56 99 90 67 83

```
07 63 87 79 29   03 06 11 80 72   96 20 74 41 56   23 82 19 95 38
60 52 88 34 41   07 95 41 98 14   59 17 52 06 95   05 53 35 21 39
83 59 63 56 55   06 95 89 29 83   05 12 80 97 19   77 43 35 37 83
10 85 06 27 46   99 59 91 05 07   13 49 90 63 19   53 07 57 18 39
39 82 09 89 52   43 62 26 31 47   64 42 18 08 14   43 80 00 93 51

59 58 00 64 78   75 56 97 88 00   88 83 55 44 86   23 76 80 61 56
38 50 80 73 41   23 79 34 87 63   90 82 29 70 22   17 71 90 42 07
30 69 27 06 68   94 68 81 61 27   56 19 68 00 91   82 06 76 34 00
65 44 39 56 59   18 28 82 74 37   49 63 22 40 41   08 33 76 56 76
27 26 75 02 64   13 19 27 22 94   07 47 74 46 06   17 98 54 89 11

91 30 70 69 91   19 07 22 42 10   36 69 95 37 28   28 82 53 57 93
68 43 49 46 88   84 47 31 36 22   62 12 69 84 08   12 84 38 25 90
48 90 81 58 77   54 74 52 45 91   35 70 00 47 54   83 82 45 26 92
06 91 34 51 97   42 67 27 86 01   11 88 30 95 28   63 01 19 89 01
10 45 51 60 19   14 21 03 37 12   91 34 23 78 21   88 32 58 08 51

12 88 39 73 43   65 02 76 11 84   04 28 50 13 92   17 97 41 50 77
21 77 83 09 76   38 80 73 69 61   31 64 94 20 96   63 28 10 20 23
19 52 35 95 15   65 12 25 96 59   86 28 36 82 58   69 57 21 37 98
67 24 55 26 70   35 58 31 65 63   79 24 68 66 86   76 46 33 42 22
60 58 44 73 77   07 50 03 79 92   45 13 42 65 29   26 76 08 36 37

53 85 34 13 77   36 06 69 48 50   58 83 87 38 59   49 36 47 33 31
24 63 73 87 36   74 38 48 93 42   52 62 30 79 92   12 36 91 86 01
83 08 01 24 51   38 99 22 28 15   07 75 95 17 77   97 37 72 75 85
16 44 42 43 34   36 15 19 90 73   27 49 37 09 39   85 13 03 25 52
60 79 01 81 57   57 17 86 57 62   11 16 17 85 76   45 81 95 29 79
```

From W. J. Dixon and F. J. Massey, *Introduction to Statistical Analysis*, 2nd edition. New York: McGraw-Hill, 1957. Reprinted by permission of McGraw-Hill and the Rand Corporation.

TABLE III Critical Values of t

For any given df, the table shows the values of t corresponding to various levels of probability. Obtained t is significant at a given level if it is equal to or greater than the value shown in the table.

df	Level of significance for one-tailed test					
	.10	.05	.025	.01	.005	.0005
	Level of significance for two-tailed test					
	.20	.10	.05	.02	.01	.001
1	3.078	6.314	12.706	31.821	63.657	636.619
2	1.886	2.920	4.303	6.965	9.925	31.598
3	1.638	2.353	3.182	4.541	5.841	12.941
4	1.533	2.132	2.776	3.747	4.604	8.610
5	1.476	2.015	2.571	3.365	4.032	6.859
6	1.440	1.943	2.447	3.143	3.707	5.959
7	1.415	1.895	2.365	2.998	3.499	5.405
8	1.397	1.860	2.306	2.896	3.355	5.041
9	1.383	1.833	2.262	2.821	3.250	4.781
10	1.372	1.812	2.228	2.764	3.169	4.587
11	1.363	1.796	2.201	2.718	3.106	4.437
12	1.356	1.782	2.179	2.681	3.055	4.318
13	1.350	1.771	2.160	2.650	3.012	4.221
14	1.345	1.761	2.145	2.624	2.977	4.140
15	1.341	1.753	2.131	2.602	2.947	4.073
16	1.337	1.746	2.120	2.583	2.921	4.015
17	1.333	1.740	2.110	2.567	2.898	3.965
18	1.330	1.734	2.101	2.552	2.878	3.922
19	1.328	1.729	2.093	2.539	2.861	3.883
20	1.325	1.725	2.086	2.528	2.845	3.850
21	1.323	1.721	2.080	2.518	2.831	3.819
22	1.321	1.717	2.074	2.508	2.819	3.792
23	1.319	1.714	2.069	2.500	2.807	3.767
24	1.318	1.711	2.064	2.492	2.797	3.745
25	1.316	1.708	2.060	2.485	2.787	3.725
26	1.315	1.706	2.056	2.479	2.779	3.707
27	1.314	1.703	2.052	2.473	2.771	3.690
28	1.313	1.701	2.048	2.467	2.763	3.674
29	1.311	1.699	2.045	2.462	2.756	3.659
30	1.310	1.697	2.042	2.457	2.750	3.646
40	1.303	1.684	2.021	2.423	2.704	3.551
60	1.296	1.671	2.000	2.390	2.660	3.460
120	1.289	1.658	1.980	2.358	2.617	3.373
∞	1.282	1.645	1.960	2.326	2.576	3.291

From R. A. Fisher and F. Yates, *Statistical Tables for Biological, Agricultural and Medical Research*. Edinburgh: Oliver and Boyd, Ltd., 1948. Reprinted by permission of the authors and publisher.

TABLE IV Normal Curve Areas

Z	.00	.01	.02	.03	.04	.05	.06	.07	.08	.09
0.0	.0000	.0040	.0080	.0120	.0160	.0199	.0239	.0279	.0319	.0359
0.1	.0398	.0438	.0478	.0517	.0557	.0596	.0636	.0675	.0714	.0753
0.2	.0793	.0832	.0871	.0910	.0948	.0987	.1026	.1064	.1103	.1141
0.3	.1179	.1217	.1255	.1293	.1331	.1368	.1406	.1443	.1480	.1517
0.4	.1554	.1591	.1628	.1664	.1700	.1736	.1772	.1808	.1844	.1879
0.5	.1915	.1950	.1985	.2019	.2054	.2088	.2123	.2157	.2190	.2224
0.6	.2257	.2291	.2324	.2357	.2389	.2422	.2454	.2486	.2517	.2549
0.7	.2580	.2611	.2642	.2673	.2704	.2734	.2764	.2794	.2823	.2852
0.8	.2881	.2910	.2939	.2967	.2995	.3023	.3051	.3078	.3106	.3133
0.9	.3159	.3186	.3212	.3238	.3264	.3289	.3315	.3340	.3365	.3389
1.0	.3413	.3438	.3461	.3485	.3508	.3531	.3554	.3577	.3599	.3621
1.1	.3643	.3665	.3686	.3708	.3729	.3749	.3770	.3790	.3810	.3830
1.2	.3849	.3869	.3888	.3907	.3925	.3944	.3962	.3980	.3997	.4015
1.3	.4032	.4049	.4066	.4082	.4099	.4115	.4131	.4147	.4162	.4177
1.4	.4192	.4207	.4222	.4236	.4251	.4265	.4279	.4292	.4306	.4319
1.5	.4332	.4345	.4357	.4370	.4382	.4394	.4406	.4418	.4429	.4441
1.6	.4452	.4463	.4474	.4484	.4495	.4505	.4515	.4525	.4535	.4545
1.7	.4554	.4564	.4573	.4582	.4591	.4599	.4608	.4616	.4625	.4633
1.8	.4641	.4649	.4656	.4664	.4671	.4678	.4686	.4693	.4699	.4706
1.9	.4713	.4719	.4726	.4732	.4738	.4744	.4750	.4756	.4761	.4767
2.0	.4772	.4778	.4783	.4788	.4793	.4798	.4803	.4808	.4812	.4817
2.1	.4821	.4826	.4830	.4834	.4838	.4842	.4846	.4850	.4854	.4857
2.2	.4861	.4864	.4868	.4871	.4875	.4878	.4881	.4884	.4887	.4890
2.3	.4893	.4896	.4898	.4901	.4904	.4906	.4909	.4911	.4913	.4916
2.4	.4918	.4920	.4922	.4925	.4927	.4929	.4931	.4932	.4934	.4936
2.5	.4938	.4940	.4941	.4943	.4945	.4946	.4948	.4949	.4951	.4952
2.6	.4953	.4955	.4956	.4957	.4959	.4960	.4961	.4962	.4963	.4964
2.7	.4965	.4966	.4967	.4968	.4969	.4970	.4971	.4972	.4973	.4974
2.8	.4974	.4975	.4976	.4977	.4977	.4978	.4979	.4979	.4980	.4981
2.9	.4981	.4982	.4982	.4983	.4984	.4984	.4985	.4985	.4986	.4986
3.0	.4987	.4987	.4987	.4988	.4988	.4989	.4989	.4989	.4990	.4990
3.1	.49903									
3.2	.49931									
3.3	.49952									
3.4	.49966									
3.5	.49977									
3.6	.49984									
3.7	.49989									
3.8	.49993									
3.9	.49995									
4.0	.50000									

TABLE V Chi Square

Column headings indicate probability of chance

deviation between O and E.

D.F. \ P	0.25	0.10	0.05	0.025	0.01	0.005
1.	1.323	2.706	3.841	5.024	6.635	7.879
2.	2.773	4.605	5.991	7.378	9.210	10.597
3.	4.108	6.251	7.815	9.348	11.345	12.838
4.	5.385	7.779	9.488	11.143	13.277	14.860
5.	6.626	9.236	11.071	12.833	15.086	16.750
6.	7.841	10.645	12.592	14.449	16.812	18.548
7.	9.037	12.017	14.067	16.013	18.475	20.278
8.	10.219	13.362	15.507	17.535	20.090	21.955
9.	11.389	14.684	16.919	19.023	21.666	23.589
10.	12.549	15.987	18.307	20.483	23.209	25.188
11.	13.701	17.275	19.675	21.920	24.725	26.757
12.	14.845	18.549	21.026	23.337	26.217	28.299
13.	15.984	19.812	22.362	24.736	27.688	29.819
14.	17.117	21.064	23.685	26.119	29.141	31.319
15.	18.245	22.307	24.996	27.488	30.578	32.801

Adapted from table of χ^2 appearing in *Handbook of Statistical Tables* by D. B. Owen, Addison-Wesley, 1962, p. 50. Reprinted by permission of the U.S. Atomic Energy Commission.

TABLE VI Values of the Exponential Function $e^{-\lambda}$

λ	0.00	0.01	0.02	0.03	0.04	0.05	0.06	0.07	0.08	0.09
0.00	1.000	0.990	0.980	0.970	0.961	0.951	0.942	0.932	0.923	0.914
0.10	0.905	0.896	0.887	0.878	0.869	0.861	0.852	0.844	0.835	0.827
0.20	0.819	0.811	0.803	0.795	0.787	0.779	0.771	0.763	0.756	0.748
0.30	0.741	0.733	0.726	0.719	0.712	0.705	0.698	0.691	0.684	0.677
0.40	0.670	0.664	0.657	0.651	0.644	0.638	0.631	0.625	0.619	0.613
0.50	0.607	0.600	0.595	0.589	0.583	0.577	0.571	0.566	0.560	0.554
0.60	0.549	0.543	0.538	0.533	0.527	0.522	0.517	0.512	0.507	0.502
0.70	0.497	0.492	0.487	0.482	0.477	0.472	0.468	0.463	0.458	0.454
0.80	0.449	0.445	0.440	0.436	0.432	0.427	0.423	0.419	0.415	0.411
0.90	0.407	0.403	0.399	0.395	0.391	0.387	0.383	0.379	0.375	0.372
1.00	0.368	0.364	0.361	0.357	0.353	0.350	0.346	0.343	0.340	0.336
1.10	0.333	0.330	0.326	0.323	0.320	0.317	0.313	0.310	0.307	0.304
1.20	0.301	0.298	0.295	0.292	0.289	0.287	0.284	0.281	0.278	0.275
1.30	0.273	0.270	0.267	0.264	0.262	0.259	0.257	0.254	0.252	0.249
1.40	0.247	0.244	0.242	0.239	0.237	0.235	0.232	0.230	0.228	0.225
1.50	0.223	0.221	0.219	0.217	0.214	0.212	0.210	0.208	0.206	0.204
1.60	0.202	0.200	0.198	0.196	0.194	0.192	0.190	0.188	0.186	0.185
1.70	0.183	0.181	0.179	0.177	0.176	0.174	0.172	0.170	0.169	0.167
1.80	0.165	0.164	0.162	0.160	0.159	0.157	0.156	0.154	0.153	0.151
1.90	0.150	0.148	0.147	0.145	0.144	0.142	0.141	0.139	0.138	0.137
2.00	0.135	0.134	0.133	0.131	0.130	0.129	0.127	0.126	0.125	0.124
2.10	0.122	0.121	0.120	0.119	0.118	0.116	0.115	0.114	0.113	0.112
2.20	0.111	0.110	0.109	0.108	0.106	0.105	0.104	0.103	0.102	0.101
2.30	0.100	0.0992	0.0983	0.0973	0.0963	0.0953	0.0944	0.0935	0.0926	0.0916
2.40	0.0907	0.0898	0.0889	0.0880	0.0872	0.0863	0.0854	0.0846	0.0837	0.0829
2.50	0.0821	0.0813	0.0805	0.0797	0.0789	0.0781	0.0773	0.0765	0.0758	0.0750
2.60	0.0743	0.0735	0.0728	0.0721	0.0714	0.0707	0.0699	0.0693	0.0686	0.0679
2.70	0.0672	0.0665	0.0659	0.0652	0.0646	0.0639	0.0633	0.0627	0.0620	0.0614
2.80	0.0608	0.0602	0.0596	0.0590	0.0584	0.0578	0.0573	0.0567	0.0561	0.0556
2.90	0.0550	0.0545	0.0539	0.0534	0.0529	0.0523	0.0518	0.0513	0.0508	0.0503

From Avram Goldstein, *Biostatistics*. New York: Macmillan, 1964. Reprinted by permission.

TABLE VII Critical Values of F

The obtained F is significant at a given level if it is equal to or greater than the value shown in the table. 0.05 (light row) and 0.01 (dark row) points for the distribution of F

Degrees of freedom for greater mean square

Each cell shows the 0.05 (light) value over the 0.01 (dark) value as "light / dark".

n_2 \ n_1	1	2	3	4	5	6	7	8	9	10	11	12	14	16	20	24	30	40	50	75	100	200	500	∞
1	161 / 4052	200 / 4999	216 / 5403	225 / 5625	230 / 5764	234 / 5859	237 / 5928	239 / 5981	241 / 6022	242 / 6056	243 / 6082	244 / 6106	245 / 6142	246 / 6169	248 / 6208	249 / 6234	250 / 6258	251 / 6286	252 / 6302	253 / 6323	253 / 6334	254 / 6352	254 / 6361	254 / 6366
2	18.51 / 98.49	19.00 / 99.01	19.16 / 99.17	19.25 / 99.25	19.30 / 99.30	19.33 / 99.33	19.36 / 99.34	19.37 / 99.36	19.38 / 99.38	19.39 / 99.40	19.40 / 99.41	19.41 / 99.42	19.42 / 99.43	19.43 / 99.44	19.44 / 99.45	19.45 / 99.46	19.46 / 99.47	19.47 / 99.48	19.47 / 99.48	19.48 / 99.49	19.49 / 99.49	19.49 / 99.49	19.50 / 99.50	19.50 / 99.50
3	10.13 / 34.12	9.55 / 30.81	9.28 / 29.46	9.12 / 28.71	9.01 / 28.24	8.94 / 27.91	8.88 / 27.67	8.84 / 27.49	8.81 / 27.34	8.78 / 27.23	8.76 / 27.13	8.74 / 27.05	8.71 / 26.92	8.69 / 26.83	8.66 / 26.69	8.64 / 26.60	8.62 / 26.50	8.60 / 26.41	8.58 / 26.30	8.57 / 26.27	8.56 / 26.23	8.54 / 26.18	8.54 / 26.14	8.53 / 26.12
4	7.71 / 21.20	6.94 / 18.00	6.59 / 16.69	6.39 / 15.98	6.26 / 15.52	6.16 / 15.21	6.09 / 14.98	6.04 / 14.80	6.00 / 14.66	5.96 / 14.54	5.93 / 14.45	5.91 / 14.37	5.87 / 14.24	5.84 / 14.15	5.80 / 14.02	5.77 / 13.93	5.74 / 13.83	5.71 / 13.74	5.70 / 13.69	5.68 / 13.61	5.66 / 13.57	5.65 / 13.52	5.64 / 13.48	5.63 / 13.46
5	6.61 / 16.26	5.79 / 13.27	5.41 / 12.06	5.19 / 11.39	5.05 / 10.97	4.95 / 10.67	4.88 / 10.45	4.82 / 10.27	4.78 / 10.15	4.74 / 10.05	4.70 / 9.96	4.68 / 9.89	4.64 / 9.77	4.60 / 9.68	4.56 / 9.55	4.53 / 9.47	4.50 / 9.38	4.46 / 9.29	4.44 / 9.24	4.42 / 9.17	4.40 / 9.13	4.38 / 9.07	4.37 / 9.04	4.36 / 9.02
6	5.99 / 13.74	5.14 / 10.92	4.76 / 9.78	4.53 / 9.15	4.39 / 8.75	4.28 / 8.47	4.21 / 8.26	4.15 / 8.10	4.10 / 7.98	4.06 / 7.87	4.03 / 7.79	4.00 / 7.72	3.96 / 7.60	3.92 / 7.52	3.87 / 7.39	3.84 / 7.31	3.81 / 7.23	3.77 / 7.14	3.75 / 7.09	3.72 / 7.02	3.71 / 6.99	3.69 / 6.94	3.68 / 6.90	3.67 / 6.88
7	5.59 / 12.25	4.74 / 9.55	4.35 / 8.45	4.12 / 7.85	3.97 / 7.46	3.87 / 7.19	3.79 / 7.00	3.73 / 6.84	3.68 / 6.71	3.63 / 6.62	3.60 / 6.54	3.57 / 6.47	3.52 / 6.35	3.49 / 6.27	3.44 / 6.15	3.41 / 6.07	3.38 / 5.98	3.34 / 5.90	3.32 / 5.85	3.29 / 5.78	3.28 / 5.75	3.25 / 5.70	3.24 / 5.67	3.23 / 5.65
8	5.32 / 11.26	4.46 / 8.65	4.07 / 7.59	3.84 / 7.01	3.69 / 6.63	3.58 / 6.37	3.50 / 6.19	3.44 / 6.03	3.39 / 5.91	3.34 / 5.82	3.31 / 5.74	3.28 / 5.67	3.23 / 5.56	3.20 / 5.48	3.15 / 5.36	3.12 / 5.28	3.08 / 5.20	3.05 / 5.11	3.03 / 5.06	3.00 / 5.00	2.98 / 4.96	2.96 / 4.91	2.94 / 4.88	2.93 / 4.86
9	5.12 / 10.56	4.26 / 8.02	3.86 / 6.99	3.63 / 6.42	3.48 / 6.06	3.37 / 5.80	3.29 / 5.62	3.23 / 5.47	3.18 / 5.35	3.13 / 5.26	3.10 / 5.18	3.07 / 5.11	3.02 / 5.00	2.98 / 4.92	2.93 / 4.80	2.90 / 4.73	2.86 / 4.64	2.82 / 4.56	2.80 / 4.51	2.77 / 4.45	2.76 / 4.41	2.73 / 4.36	2.72 / 4.33	2.71 / 4.31
10	4.96 / 10.04	4.10 / 7.56	3.71 / 6.55	3.48 / 5.99	3.33 / 5.64	3.22 / 5.39	3.14 / 5.21	3.07 / 5.06	3.02 / 4.95	2.97 / 4.85	2.94 / 4.78	2.91 / 4.71	2.86 / 4.60	2.82 / 4.52	2.77 / 4.41	2.74 / 4.33	2.70 / 4.25	2.67 / 4.17	2.64 / 4.12	2.61 / 4.05	2.59 / 4.01	2.56 / 3.96	2.55 / 3.93	2.54 / 3.91
11	4.84 / 9.65	3.98 / 7.20	3.59 / 6.22	3.36 / 5.67	3.20 / 5.32	3.09 / 5.07	3.01 / 4.88	2.95 / 4.74	2.90 / 4.63	2.86 / 4.54	2.82 / 4.46	2.79 / 4.40	2.74 / 4.29	2.70 / 4.21	2.65 / 4.10	2.61 / 4.02	2.57 / 3.94	2.53 / 3.86	2.50 / 3.80	2.47 / 3.74	2.45 / 3.70	2.42 / 3.66	2.41 / 3.62	2.40 / 3.60
12	4.75 / 9.33	3.88 / 6.93	3.49 / 5.95	3.26 / 5.41	3.11 / 5.06	3.00 / 4.82	2.92 / 4.65	2.85 / 4.50	2.80 / 4.39	2.76 / 4.30	2.72 / 4.22	2.69 / 4.16	2.64 / 4.05	2.60 / 3.98	2.54 / 3.86	2.50 / 3.78	2.46 / 3.70	2.42 / 3.61	2.40 / 3.56	2.36 / 3.49	2.35 / 3.46	2.32 / 3.41	2.31 / 3.38	2.30 / 3.36
13	4.67 / 9.07	3.80 / 6.70	3.41 / 5.74	3.18 / 5.20	3.02 / 4.86	2.92 / 4.62	2.84 / 4.44	2.77 / 4.30	2.72 / 4.19	2.67 / 4.10	2.63 / 4.02	2.60 / 3.96	2.55 / 3.85	2.51 / 3.78	2.46 / 3.67	2.42 / 3.59	2.38 / 3.51	2.34 / 3.42	2.32 / 3.37	2.28 / 3.30	2.26 / 3.27	2.24 / 3.21	2.22 / 3.18	2.21 / 3.16
14	4.60 / 8.86	3.74 / 6.51	3.34 / 5.56	3.11 / 5.03	2.96 / 4.69	2.85 / 4.46	2.77 / 4.28	2.70 / 4.14	2.65 / 4.03	2.60 / 3.94	2.56 / 3.86	2.53 / 3.80	2.48 / 3.70	2.44 / 3.62	2.39 / 3.51	2.35 / 3.43	2.31 / 3.34	2.27 / 3.26	2.24 / 3.21	2.21 / 3.14	2.19 / 3.11	2.16 / 3.06	2.14 / 3.02	2.13 / 3.00
15	4.54 / 8.68	3.68 / 6.36	3.29 / 5.42	3.06 / 4.89	2.90 / 4.56	2.79 / 4.32	2.70 / 4.14	2.64 / 4.00	2.59 / 3.89	2.55 / 3.80	2.51 / 3.73	2.48 / 3.67	2.43 / 3.56	2.39 / 3.48	2.33 / 3.36	2.29 / 3.29	2.25 / 3.20	2.21 / 3.12	2.18 / 3.07	2.15 / 3.00	2.12 / 2.97	2.10 / 2.92	2.08 / 2.89	2.07 / 2.87

Degrees of freedom for lesser mean square

A–10

0.05 (light row) and 0.01 (dark row) points for the distribution of F

Degrees of freedom for greater mean square

	1	2	3	4	5	6	7	8	9	10	11	12	14	16	20	24	30	40	50	75	100	200	500	∞
16	4.49 8.53	3.63 6.23	3.24 5.29	3.01 4.77	2.85 4.44	2.74 4.20	2.66 4.03	2.59 3.89	2.54 3.78	2.49 3.69	2.45 3.61	2.42 3.55	2.37 3.45	2.33 3.37	2.28 3.25	2.24 3.18	2.20 3.10	2.16 3.01	2.13 2.96	2.09 2.89	2.07 2.86	2.04 2.80	2.02 2.77	2.01 2.75
17	4.45 8.40	3.59 6.11	3.20 5.18	2.96 4.67	2.81 4.34	2.70 4.10	2.62 3.93	2.55 3.79	2.50 3.68	2.45 3.59	2.41 3.52	2.38 3.45	2.33 3.35	2.29 3.27	2.23 3.16	2.19 3.08	2.15 3.00	2.11 2.92	2.08 2.86	2.04 2.79	2.02 2.76	1.99 2.70	1.97 2.67	1.96 2.65
18	4.41 8.28	3.55 6.01	3.16 5.09	2.93 4.58	2.77 4.25	2.66 4.01	2.58 3.85	2.51 3.71	2.46 3.60	2.41 3.51	2.37 3.44	2.34 3.37	2.29 3.27	2.25 3.19	2.19 3.07	2.15 3.00	2.11 2.91	2.07 2.83	2.04 2.78	2.00 2.71	1.98 2.68	1.95 2.62	1.93 2.59	1.92 2.57
19	4.38 8.18	3.52 5.93	3.13 5.01	2.90 4.50	2.74 4.17	2.63 3.94	2.55 3.77	2.48 3.63	2.43 3.52	2.38 3.43	2.34 3.36	2.31 3.30	2.26 3.19	2.21 3.12	2.15 3.00	2.11 2.92	2.07 2.84	2.02 2.76	2.00 2.70	1.96 2.63	1.94 2.60	1.91 2.54	1.90 2.51	1.88 2.49
20	4.35 8.10	3.49 5.85	3.10 4.94	2.87 4.43	2.71 4.10	2.60 3.87	2.52 3.71	2.45 3.56	2.40 3.45	2.35 3.37	2.31 3.30	2.28 3.23	2.23 3.13	2.18 3.05	2.12 2.94	2.08 2.86	2.04 2.77	1.99 2.69	1.96 2.63	1.92 2.56	1.90 2.53	1.87 2.47	1.85 2.44	1.84 2.42
21	4.32 8.02	3.47 5.78	3.07 4.87	2.84 4.37	2.68 4.04	2.57 3.81	2.49 3.65	2.42 3.51	2.37 3.40	2.32 3.31	2.28 3.24	2.25 3.17	2.20 3.07	2.15 2.99	2.09 2.88	2.05 2.80	2.00 2.72	1.96 2.63	1.93 2.58	1.89 2.51	1.87 2.47	1.84 2.42	1.82 2.38	1.81 2.36
22	4.30 7.94	3.44 5.72	3.05 4.82	2.82 4.31	2.66 3.99	2.55 3.76	2.47 3.59	2.40 3.45	2.35 3.35	2.30 3.26	2.26 3.18	2.23 3.12	2.18 3.02	2.13 2.94	2.07 2.83	2.03 2.75	1.98 2.67	1.93 2.58	1.91 2.53	1.87 2.46	1.84 2.42	1.81 2.37	1.80 2.33	1.78 2.31
23	4.28 7.88	3.42 5.66	3.03 4.76	2.80 4.26	2.64 3.94	2.53 3.71	2.45 3.54	2.38 3.41	2.32 3.30	2.28 3.21	2.24 3.14	2.20 3.07	2.14 2.97	2.10 2.89	2.04 2.78	2.00 2.70	1.96 2.62	1.91 2.53	1.88 2.48	1.84 2.41	1.82 2.37	1.79 2.32	1.77 2.28	1.76 2.26
24	4.26 7.82	3.40 5.61	3.01 4.72	2.78 4.22	2.62 3.90	2.51 3.67	2.43 3.50	2.36 3.36	2.30 3.25	2.26 3.17	2.22 3.09	2.18 3.03	2.13 2.93	2.09 2.85	2.02 2.74	1.98 2.66	1.94 2.58	1.89 2.49	1.86 2.44	1.82 2.36	1.80 2.33	1.76 2.27	1.74 2.23	1.73 2.21
25	4.24 7.77	3.38 5.57	2.99 4.68	2.76 4.18	2.60 3.86	2.49 3.63	2.41 3.46	2.34 3.32	2.28 3.21	2.24 3.13	2.20 3.05	2.16 2.99	2.11 2.89	2.06 2.81	2.00 2.70	1.96 2.62	1.92 2.54	1.87 2.45	1.84 2.40	1.80 2.32	1.77 2.29	1.74 2.23	1.72 2.19	1.71 2.17
26	4.22 7.72	3.37 5.53	2.98 4.64	2.74 4.14	2.59 3.82	2.47 3.59	2.39 3.42	2.32 3.29	2.27 3.17	2.22 3.09	2.18 3.02	2.15 2.96	2.10 2.86	2.05 2.77	1.99 2.66	1.95 2.58	1.90 2.50	1.85 2.41	1.82 2.36	1.78 2.28	1.76 2.25	1.72 2.19	1.70 2.15	1.69 2.13
27	4.21 7.68	3.35 5.49	2.96 4.60	2.73 4.11	2.57 3.79	2.46 3.56	2.37 3.39	2.30 3.26	2.25 3.14	2.20 3.06	2.16 2.98	2.13 2.93	2.08 2.83	2.03 2.74	1.97 2.63	1.93 2.55	1.88 2.47	1.84 2.38	1.80 2.33	1.76 2.25	1.74 2.21	1.71 2.16	1.68 2.12	1.67 2.10
28	4.20 7.64	3.34 5.45	2.95 4.57	2.71 4.07	2.56 3.76	2.44 3.53	2.36 3.36	2.29 3.23	3.24 3.11	2.19 3.03	2.15 2.95	2.12 2.90	2.06 2.80	2.02 2.71	1.96 2.60	1.91 2.52	1.87 2.44	1.81 2.35	1.78 2.30	1.75 2.22	1.72 2.18	1.69 2.13	1.67 2.09	1.65 2.06
29	4.18 7.60	3.33 5.42	2.93 4.54	2.70 4.04	2.54 3.73	2.43 3.50	2.35 3.33	2.28 3.20	2.22 3.08	2.18 3.00	2.14 2.92	2.10 2.87	2.05 2.77	2.00 2.68	1.94 2.57	1.90 2.49	1.85 2.41	1.80 2.32	1.77 2.27	1.73 2.19	1.71 2.15	1.68 2.10	1.65 2.06	1.64 2.03
30	4.17 7.56	3.32 5.39	2.92 4.51	2.69 4.02	2.53 3.70	2.42 3.47	2.34 3.30	2.27 3.17	2.21 3.06	2.16 2.98	2.12 2.90	2.09 2.84	2.04 2.74	1.99 2.66	1.93 2.55	1.89 2.47	1.84 2.38	1.79 2.29	1.76 2.24	1.72 2.16	1.69 2.13	1.66 2.07	1.64 2.03	1.62 2.01

Degrees of freedom for lesser mean square

(continued)

TABLE VII (*continued*)

0.05 (light row) and 0.01 (dark row) points for the distribution of F

Degrees of freedom for greater mean square

df (lesser)	1	2	3	4	5	6	7	8	9	10	11	12	14	16	20	24	30	40	50	75	100	200	500	∞
32	4.15 / 7.50	3.30 / 5.34	2.90 / 4.46	2.67 / 3.97	2.51 / 3.66	2.40 / 3.42	2.32 / 3.25	2.25 / 3.12	2.19 / 3.01	2.14 / 2.94	2.10 / 2.86	2.07 / 2.80	2.02 / 2.70	1.97 / 2.62	1.91 / 2.51	1.86 / 2.42	1.82 / 2.34	1.76 / 2.25	1.74 / 2.20	1.69 / 2.12	1.67 / 2.08	1.64 / 2.02	1.61 / 1.98	1.59 / 1.96
34	4.13 / 7.44	3.28 / 5.29	2.88 / 4.42	2.65 / 3.93	2.49 / 3.61	2.38 / 3.38	2.30 / 3.21	2.23 / 3.08	2.17 / 2.97	2.12 / 2.89	2.08 / 2.82	2.05 / 2.76	2.00 / 2.66	1.95 / 2.58	1.89 / 2.47	1.84 / 2.38	1.80 / 2.30	1.74 / 2.21	1.71 / 2.15	1.67 / 2.08	1.64 / 2.04	1.61 / 1.98	1.59 / 1.94	1.57 / 1.91
36	4.11 / 7.39	3.26 / 5.25	2.86 / 4.38	2.63 / 3.89	2.48 / 3.58	2.36 / 3.35	2.28 / 3.18	2.21 / 3.04	2.15 / 2.94	2.10 / 2.86	2.06 / 2.78	2.03 / 2.72	1.98 / 2.62	1.93 / 2.54	1.87 / 2.43	1.82 / 2.35	1.78 / 2.26	1.72 / 2.17	1.69 / 2.12	1.65 / 2.04	1.62 / 2.00	1.59 / 1.94	1.56 / 1.90	1.55 / 1.87
38	4.10 / 7.35	3.25 / 5.21	2.85 / 4.34	2.62 / 3.86	2.46 / 3.54	2.35 / 3.32	2.26 / 3.15	2.19 / 3.02	2.14 / 2.91	2.09 / 2.82	2.05 / 2.75	2.02 / 2.69	1.96 / 2.59	1.92 / 2.51	1.85 / 2.40	1.80 / 2.32	1.76 / 2.22	1.71 / 2.14	1.67 / 2.08	1.63 / 2.00	1.60 / 1.97	1.57 / 1.90	1.54 / 1.86	1.53 / 1.84
40	4.08 / 7.31	3.23 / 5.18	2.84 / 4.31	2.61 / 3.83	2.45 / 3.51	2.34 / 3.29	2.25 / 3.12	2.18 / 2.99	2.12 / 2.88	2.07 / 2.80	2.04 / 2.73	2.00 / 2.66	1.95 / 2.56	1.90 / 2.49	1.84 / 2.37	1.79 / 2.29	1.74 / 2.20	1.69 / 2.11	1.66 / 2.05	1.61 / 1.97	1.59 / 1.94	1.55 / 1.88	1.53 / 1.84	1.51 / 1.81
42	4.07 / 7.27	3.22 / 5.15	2.83 / 4.29	2.59 / 3.80	2.44 / 3.49	2.32 / 3.26	2.24 / 3.10	2.17 / 2.96	2.11 / 2.86	2.06 / 2.77	2.02 / 2.70	1.99 / 2.64	1.94 / 2.54	1.89 / 2.46	1.82 / 2.35	1.78 / 2.26	1.73 / 2.17	1.68 / 2.08	1.64 / 2.02	1.60 / 1.94	1.57 / 1.91	1.54 / 1.85	1.51 / 1.80	1.49 / 1.78
44	4.06 / 7.24	3.21 / 5.12	2.82 / 4.26	2.58 / 3.78	2.43 / 3.46	2.31 / 3.24	2.23 / 3.07	2.16 / 2.94	2.10 / 2.84	2.05 / 2.75	2.01 / 2.68	1.98 / 2.62	1.92 / 2.52	1.88 / 2.44	1.81 / 2.32	1.76 / 2.24	1.72 / 2.15	1.66 / 2.06	1.63 / 2.00	1.58 / 1.92	1.56 / 1.88	1.52 / 1.82	1.50 / 1.78	1.48 / 1.75
46	4.05 / 7.21	3.20 / 5.10	2.81 / 4.24	2.57 / 3.76	2.42 / 3.44	2.30 / 3.22	2.22 / 3.05	2.14 / 2.92	2.09 / 2.82	2.04 / 2.73	2.00 / 2.66	1.97 / 2.60	1.91 / 2.50	1.87 / 2.42	1.80 / 2.30	1.75 / 2.22	1.71 / 2.13	1.65 / 2.04	1.62 / 1.98	1.57 / 1.90	1.54 / 1.86	1.51 / 1.80	1.48 / 1.76	1.46 / 1.72
48	4.04 / 7.19	3.19 / 5.08	2.80 / 4.22	2.56 / 3.74	2.41 / 3.42	2.30 / 3.20	2.21 / 3.04	2.14 / 2.90	2.08 / 2.80	2.03 / 2.71	1.99 / 2.64	1.96 / 2.58	1.90 / 2.48	1.86 / 2.40	1.79 / 2.28	1.74 / 2.20	1.70 / 2.11	1.64 / 2.02	1.61 / 1.96	1.56 / 1.88	1.53 / 1.84	1.50 / 1.78	1.47 / 1.73	1.45 / 1.70
50	4.03 / 7.17	3.18 / 5.06	2.79 / 4.20	2.56 / 3.72	2.40 / 3.41	2.29 / 3.18	2.20 / 3.02	2.13 / 2.88	2.07 / 2.78	2.02 / 2.70	1.98 / 2.62	1.95 / 2.56	1.90 / 2.46	1.85 / 2.39	1.78 / 2.26	1.74 / 2.18	1.69 / 2.10	1.63 / 2.00	1.60 / 1.94	1.55 / 1.86	1.52 / 1.82	1.48 / 1.76	1.46 / 1.71	1.44 / 1.68
55	4.02 / 7.12	3.17 / 5.01	2.78 / 4.16	2.54 / 3.68	2.38 / 3.37	2.27 / 3.15	2.18 / 2.98	2.11 / 2.85	2.05 / 2.75	2.00 / 2.66	1.97 / 2.59	1.93 / 2.53	1.88 / 2.43	1.83 / 2.35	1.76 / 2.23	1.72 / 2.15	1.67 / 2.06	1.61 / 1.96	1.58 / 1.90	1.52 / 1.82	1.50 / 1.78	1.46 / 1.71	1.43 / 1.66	1.41 / 1.64
60	4.00 / 7.08	3.15 / 4.98	2.76 / 4.13	2.52 / 3.65	2.37 / 3.34	2.25 / 3.12	2.17 / 2.95	2.10 / 2.82	2.04 / 2.72	1.99 / 2.63	1.95 / 2.56	1.92 / 2.50	1.86 / 2.40	1.81 / 2.32	1.75 / 2.20	1.70 / 2.12	1.65 / 2.03	1.59 / 1.93	1.56 / 1.87	1.50 / 1.79	1.48 / 1.74	1.44 / 1.68	1.41 / 1.63	1.39 / 1.60
65	3.99 / 7.04	3.14 / 4.95	2.75 / 4.10	2.51 / 3.62	2.36 / 3.31	2.24 / 3.09	2.15 / 2.93	2.08 / 2.79	2.02 / 2.70	1.98 / 2.61	1.94 / 2.54	1.90 / 2.47	1.85 / 2.37	1.80 / 2.30	1.73 / 2.18	1.68 / 2.09	1.63 / 2.00	1.57 / 1.90	1.54 / 1.84	1.49 / 1.76	1.46 / 1.71	1.42 / 1.64	1.39 / 1.60	1.37 / 1.56
70	3.98 / 7.01	3.13 / 4.92	2.74 / 4.08	2.50 / 3.60	2.35 / 3.29	2.23 / 3.07	2.14 / 2.91	2.07 / 2.77	2.01 / 2.67	1.97 / 2.59	1.93 / 2.51	1.89 / 2.45	1.84 / 2.35	1.79 / 2.28	1.72 / 2.15	1.67 / 2.07	1.62 / 1.98	1.56 / 1.88	1.53 / 1.82	1.47 / 1.74	1.45 / 1.69	1.40 / 1.62	1.37 / 1.56	1.35 / 1.53
80	3.96 / 6.96	3.11 / 4.88	2.72 / 4.04	2.48 / 3.56	2.33 / 3.25	2.21 / 3.04	2.12 / 2.87	2.05 / 2.74	1.99 / 2.64	1.95 / 2.55	1.91 / 2.48	1.88 / 2.41	1.82 / 2.32	1.77 / 2.24	1.70 / 2.11	1.65 / 2.03	1.60 / 1.94	1.54 / 1.84	1.51 / 1.78	1.45 / 1.70	1.42 / 1.65	1.38 / 1.57	1.35 / 1.52	1.32 / 1.49

Degrees of freedom for lesser mean square

0.05 (light row) and 0.01 (dark row) points for the distribution of F

Degrees of freedom for greater mean square

Degrees of freedom for lesser mean square	1	2	3	4	5	6	7	8	9	10	11	12	14	16	20	24	30	40	50	75	100	200	500	∞
100	3.94 / 6.90	3.09 / 4.82	2.70 / 3.98	2.46 / 3.51	2.30 / 3.20	2.19 / 2.99	2.10 / 2.82	2.03 / 2.69	1.97 / 2.59	1.92 / 2.51	1.88 / 2.43	1.85 / 2.36	1.79 / 2.26	1.75 / 2.19	1.68 / 2.06	1.63 / 1.98	1.57 / 1.89	1.51 / 1.79	1.48 / 1.73	1.42 / 1.64	1.39 / 1.59	1.34 / 1.51	1.30 / 1.46	1.28 / 1.43
125	3.92 / 6.84	3.07 / 4.78	2.68 / 3.94	2.44 / 3.47	2.29 / 3.17	2.17 / 2.95	2.08 / 2.79	2.01 / 2.65	1.95 / 2.56	1.90 / 2.47	1.86 / 2.40	1.83 / 2.33	1.77 / 2.23	1.72 / 2.15	1.65 / 2.03	1.60 / 1.94	1.55 / 1.85	1.49 / 1.75	1.45 / 1.68	1.39 / 1.59	1.36 / 1.54	1.31 / 1.46	1.27 / 1.40	1.25 / 1.37
150	3.91 / 6.81	3.06 / 4.75	2.67 / 3.91	2.43 / 3.44	2.27 / 3.13	2.16 / 2.92	2.07 / 2.76	2.00 / 2.62	1.94 / 2.53	1.89 / 2.44	1.85 / 2.37	1.82 / 2.30	1.76 / 2.20	1.71 / 2.12	1.64 / 2.00	1.59 / 1.91	1.54 / 1.83	1.47 / 1.72	1.44 / 1.66	1.37 / 1.56	1.34 / 1.51	1.29 / 1.43	1.25 / 1.37	1.22 / 1.33
200	3.89 / 6.76	3.04 / 4.71	2.65 / 3.88	2.41 / 3.41	2.26 / 3.11	2.14 / 2.90	2.05 / 2.73	1.98 / 2.60	1.92 / 2.50	1.87 / 2.41	1.83 / 2.34	1.80 / 2.28	1.74 / 2.17	1.69 / 2.09	1.62 / 1.97	1.57 / 1.88	1.52 / 1.79	1.45 / 1.69	1.42 / 1.62	1.35 / 1.53	1.32 / 1.48	1.26 / 1.39	1.22 / 1.33	1.19 / 1.28
400	3.86 / 6.70	3.02 / 4.66	2.62 / 3.83	2.39 / 3.36	2.23 / 3.06	2.12 / 2.85	2.03 / 2.69	1.96 / 2.55	1.90 / 2.46	1.85 / 2.37	1.81 / 2.29	1.78 / 2.23	1.72 / 2.12	1.67 / 2.04	1.60 / 1.92	1.54 / 1.84	1.49 / 1.74	1.42 / 1.64	1.38 / 1.57	1.32 / 1.47	1.28 / 1.42	1.22 / 1.32	1.16 / 1.24	1.13 / 1.19
1000	3.85 / 6.66	3.00 / 4.62	2.61 / 3.80	2.38 / 3.34	2.22 / 3.04	2.10 / 2.82	2.02 / 2.66	1.95 / 2.53	1.89 / 2.43	1.84 / 2.34	1.80 / 2.26	1.76 / 2.20	1.70 / 2.09	1.65 / 2.01	1.58 / 1.89	1.53 / 1.81	1.47 / 1.71	1.41 / 1.61	1.36 / 1.54	1.30 / 1.44	1.26 / 1.38	1.19 / 1.28	1.13 / 1.19	1.08 / 1.11
∞	3.84 / 6.64	2.99 / 4.60	2.60 / 3.78	2.37 / 3.32	2.21 / 3.02	2.09 / 2.80	2.01 / 2.64	1.94 / 2.51	1.88 / 2.41	1.83 / 2.32	1.79 / 2.24	1.75 / 2.18	1.69 / 2.07	1.64 / 1.99	1.57 / 1.87	1.52 / 1.79	1.46 / 1.69	1.40 / 1.59	1.35 / 1.52	1.28 / 1.41	1.24 / 1.36	1.17 / 1.25	1.11 / 1.15	1.00 / 1.00

TABLE VIII Transformation of Percentage to Arcsin $\sqrt{\text{percentage}}$

The numbers in this table are the angles (in degrees) corresponding to given percentages under the transformation arcsin $\sqrt{\text{percentage}}$.

%	0	1	2	3	4	5	6	7	8	9
0.0	0	0.57	0.81	0.99	1.15	1.28	1.40	1.52	1.62	1.72
0.1	1.81	1.90	1.99	2.07	2.14	2.22	2.29	2.36	2.43	2.50
0.2	2.56	2.63	2.69	2.75	2.81	2.87	2.92	2.98	3.03	3.09
0.3	3.14	3.19	3.24	3.29	3.34	3.39	3.44	3.49	3.53	3.58
0.4	3.63	3.67	3.72	3.76	3.80	3.85	3.89	3.93	3.97	4.01
0.5	4.05	4.09	4.13	4.17	4.21	4.25	4.29	4.33	4.37	4.40
0.6	4.44	4.48	4.52	4.55	4.59	4.62	4.66	4.69	4.73	4.76
0.7	4.80	4.83	4.87	4.90	4.93	4.97	5.00	5.03	5.07	5.10
0.8	5.13	5.16	5.20	5.23	5.26	5.29	5.32	5.35	5.38	5.41
0.9	5.44	5.47	5.50	5.53	5.56	5.59	5.62	5.65	5.68	5.71
1	5.74	6.02	6.29	6.55	6.80	7.04	7.27	7.49	7.71	7.92
2	8.13	8.33	8.53	8.72	8.91	9.10	9.28	9.46	9.63	9.81
3	9.98	10.14	10.31	10.47	10.63	10.78	10.94	11.09	11.24	11.39
4	11.54	11.68	11.83	11.97	12.11	12.25	12.39	12.52	12.66	12.79
5	12.92	13.05	13.18	13.31	13.44	13.56	13.69	13.81	13.94	14.06
6	14.18	14.30	14.42	14.54	14.65	14.77	14.89	15.00	15.12	15.23
7	15.34	15.45	15.56	15.68	15.79	15.89	16.00	16.11	16.22	16.32
8	16.43	16.54	16.64	16.74	16.85	16.95	17.05	17.16	17.26	17.36
9	17.46	17.56	17.66	17.76	17.85	17.95	18.05	18.15	18.24	18.34
10	18.44	18.53	18.63	18.72	18.81	18.91	19.00	19.09	19.19	19.28
11	19.37	19.46	19.55	19.64	19.73	19.82	19.91	20.00	20.09	20.18
12	20.27	20.36	20.44	20.53	20.62	20.70	20.79	20.88	20.96	21.05
13	21.13	21.22	21.30	21.39	21.47	21.56	21.64	21.72	21.81	21.89
14	21.97	22.06	22.14	22.22	22.30	22.38	22.46	22.55	22.63	22.71
15	22.79	22.87	22.95	23.03	23.11	23.19	23.26	23.34	23.42	23.50
16	23.58	23.66	23.73	23.81	23.89	23.97	24.04	24.12	24.20	24.27
17	24.35	24.43	24.50	24.58	24.65	24.73	24.80	24.88	24.95	25.03
18	25.10	25.18	25.25	25.33	25.40	25.48	25.55	25.62	25.70	25.77
19	25.84	25.92	25.99	26.06	26.13	26.21	26.28	26.35	26.42	26.49
20	26.56	26.64	26.71	26.78	26.85	26.92	26.99	27.06	27.13	27.20
21	27.28	27.35	27.42	27.49	27.56	27.63	27.69	27.76	27.83	27.90
22	27.97	28.04	28.11	28.18	28.25	28.32	28.38	28.45	28.52	28.59
23	28.66	28.73	28.79	28.86	28.93	29.00	29.06	29.13	29.20	29.27
24	29.33	29.40	29.47	29.53	29.60	29.67	29.73	29.80	29.87	29.93
25	30.00	30.07	30.13	30.20	30.26	30.33	30.40	30.46	30.53	30.59
26	30.66	30.72	30.79	30.85	30.92	30.98	31.05	31.11	31.18	31.24
27	31.31	31.37	31.44	31.50	31.56	31.63	31.69	31.76	31.82	31.88
28	31.95	32.01	32.08	32.14	32.20	32.27	32.33	32.39	32.46	32.52
29	32.58	32.65	32.71	32.77	32.83	32.90	32.96	33.02	33.09	33.15
30	33.21	33.27	33.34	33.40	33.46	33.52	33.58	33.65	33.71	33.77

TABLE VIII (*continued*)

%	0	1	2	3	4	5	6	7	8	9
31	33.83	33.89	33.96	34.02	34.08	34.14	34.20	34.27	34.33	34.39
32	34.45	34.51	34.57	34.63	34.70	34.76	34.82	34.88	34.94	35.00
33	35.06	35.12	35.18	35.24	35.30	35.37	35.43	35.49	35.55	35.61
34	35.67	35.73	35.79	35.85	35.91	35.97	36.03	36.09	36.15	36.21
35	36.27	36.33	36.39	36.45	36.51	36.57	36.63	36.69	36.75	36.81
36	36.87	36.93	36.99	37.05	37.11	37.17	37.23	37.29	37.35	37.41
37	37.47	37.52	37.58	37.64	37.70	37.76	37.82	37.88	37.94	38.00
38	38.06	38.12	38.17	38.23	38.29	38.35	38.41	38.47	38.53	38.59
39	38.65	38.70	38.76	38.82	38.88	38.94	39.00	39.06	39.11	39.17
40	39.23	39.29	39.35	39.41	39.47	39.52	39.58	39.64	39.70	39.76
41	39.82	39.87	39.93	39.99	40.05	40.11	40.16	40.22	40.28	40.34
42	40.40	40.46	40.51	40.57	40.63	40.69	40.74	40.80	40.86	40.92
43	40.98	41.03	41.09	41.15	41.21	41.27	41.32	41.38	41.44	41.50
44	41.55	41.61	41.67	41.73	41.78	41.84	41.90	41.96	42.02	42.07
45	42.13	42.19	42.25	42.30	42.36	42.42	42.48	42.53	42.59	42.65
46	42.71	42.76	42.82	42.88	42.94	42.99	43.05	43.11	43.17	43.22
47	43.28	43.34	43.39	43.45	43.51	43.57	43.62	43.68	43.74	43.80
48	43.85	43.91	43.97	44.03	44.08	44.14	44.20	44.25	44.31	44.37
49	44.43	44.48	44.54	44.60	44.66	44.71	44.77	44.83	44.89	44.94
50	45.00	45.06	45.11	45.17	45.23	45.29	45.34	45.40	45.46	45.52
51	45.57	45.63	45.69	45.75	45.80	45.86	45.92	45.97	46.03	46.09
52	46.15	46.20	46.26	46.32	46.38	46.43	46.49	46.55	46.61	46.66
53	46.72	46.78	46.83	46.89	46.95	47.01	47.06	47.12	47.18	47.24
54	47.29	47.35	47.41	47.47	47.52	47.58	47.64	47.70	47.75	47.81
55	47.87	47.93	47.98	48.04	48.10	48.16	48.22	48.27	48.33	48.39
56	48.45	48.50	48.56	48.62	48.68	48.73	48.79	48.85	48.91	48.97
57	49.02	49.08	49.14	49.20	49.26	49.31	49.37	49.43	49.49	49.54
58	49.60	49.66	49.72	49.78	49.84	49.89	49.95	50.01	50.07	50.13
59	50.18	50.24	50.30	50.36	50.42	50.48	50.53	50.59	50.65	50.71
60	50.77	50.83	50.89	50.94	51.00	51.06	51.12	51.18	51.24	51.30
61	51.35	51.41	51.47	51.53	51.59	51.65	51.71	51.77	51.83	51.88
62	51.94	52.00	52.06	52.12	52.18	52.24	52.30	52.36	52.42	52.48
63	52.53	52.59	52.65	52.71	52.77	52.83	52.89	52.95	53.01	53.07
64	53.13	53.19	53.25	53.31	53.37	53.43	53.49	53.55	53.61	53.67
65	53.73	53.79	53.85	53.91	53.97	54.03	54.09	54.15	54.21	54.27
66	54.33	54.39	54.45	54.51	54.57	54.63	54.70	54.76	54.82	54.88
67	54.94	55.00	55.06	55.12	55.18	55.24	55.30	55.37	55.43	55.49
68	55.55	55.61	55.67	55.73	55.80	55.86	55.92	55.98	56.04	56.11
69	56.17	56.23	56.29	56.35	56.42	56.48	56.54	56.60	56.66	56.73
70	56.79	56.85	56.91	56.98	57.04	57.10	57.17	57.23	57.29	57.35
71	57.42	57.48	57.54	57.61	57.67	57.73	57.80	57.86	57.92	57.99
72	58.05	58.12	58.18	58.24	58.31	58.37	58.44	58.50	58.56	58.63
73	58.69	58.76	58.82	58.89	58.95	59.02	59.08	59.15	59.21	59.28
74	59.34	59.41	59.47	59.54	59.60	59.67	59.74	59.80	59.87	59.93
75	60.00	60.07	60.13	60.20	60.27	60.33	60.40	60.47	60.53	60.60

(*continued*)

TABLE VIII (*continued*)

%	0	1	2	3	4	5	6	7	8	9
76	60.67	60.73	60.80	60.87	60.94	61.00	61.07	61.14	61.21	61.27
77	61.34	61.41	61.48	61.55	61.62	61.68	61.75	61.82	61.89	61.96
78	62.03	62.10	62.17	62.24	62.31	62.37	62.44	62.51	62.58	62.65
79	62.72	62.80	62.87	62.94	63.01	63.08	63.15	63.22	63.29	63.36
80	63.44	63.51	63.58	63.65	63.72	63.79	63.87	63.94	64.01	64.08
81	64.16	64.23	64.30	64.38	64.45	64.52	64.60	64.67	64.75	64.82
82	64.90	64.97	65.05	65.12	65.20	65.27	65.35	65.42	65.50	65.57
83	65.65	65.73	65.80	65.88	65.96	66.03	66.11	66.19	66.27	66.34
84	66.42	66.50	66.58	66.66	66.74	66.81	66.89	66.97	67.05	67.13
85	67.21	67.29	67.37	67.45	67.54	67.62	67.70	67.78	67.86	67.94
86	68.03	68.11	68.19	68.28	68.36	68.44	68.53	68.61	68.70	68.78
87	68.87	68.95	69.04	69.12	69.21	69.30	69.38	69.47	69.56	69.64
88	69.73	69.82	69.91	70.00	70.09	70.18	70.27	70.36	70.45	70.54
89	70.63	70.72	70.81	70.91	71.00	71.09	71.19	71.28	71.37	71.47
90	71.56	71.66	71.76	71.85	71.95	72.05	72.15	72.24	72.34	72.44
91	72.54	72.64	72.74	72.84	72.95	73.05	73.15	73.26	73.36	73.46
92	73.57	73.68	73.78	73.89	74.00	74.11	74.21	74.32	74.44	74.55
93	74.66	74.77	74.88	75.00	75.11	75.23	75.35	75.46	75.58	75.70
94	75.82	75.94	76.06	76.19	76.31	76.44	76.56	76.69	76.82	76.95
95	77.08	77.21	77.34	77.48	77.61	77.75	77.89	78.03	78.17	78.32
96	78.46	78.61	78.76	78.91	79.06	79.22	79.37	79.53	79.69	79.86
97	80.02	80.19	80.37	80.54	80.72	80.90	81.09	81.28	81.47	81.67
98	81.87	82.08	82.29	82.51	82.73	82.96	83.20	83.45	83.71	83.98
99.0	84.26	84.29	84.32	84.35	84.38	84.41	84.44	84.47	84.50	84.53
99.1	84.56	84.59	84.62	84.65	84.68	84.71	84.74	84.77	84.80	84.84
99.2	84.87	84.90	84.93	84.97	85.00	85.03	85.07	85.10	85.13	85.17
99.3	85.20	85.24	85.27	85.31	85.34	85.38	85.41	85.45	85.48	85.52
99.4	85.56	85.60	85.63	85.67	85.71	85.75	85.79	85.83	85.87	85.91
99.5	85.95	85.99	86.03	86.07	86.11	86.15	86.20	86.24	86.28	86.33
99.6	86.37	86.42	86.47	86.51	86.56	86.61	86.66	86.71	86.76	86.81
99.7	86.86	86.91	86.97	87.02	87.08	87.13	87.19	87.25	87.31	87.37
99.8	87.44	87.50	87.57	87.64	87.71	87.78	87.86	87.93	88.01	88.10
99.9	88.19	88.28	88.38	88.48	88.60	88.72	88.85	89.01	89.19	89.43
100.0	90.00	—	—	—	—	—	—	—	—	—

Table VIII appeared in *Plant Protection* (Leningrad), Vol. 12 (1937), p. 67, and is reproduced by permission of the author, C. I. Bliss.

TABLE IX Wilcoxon Distribution—Unpaired Data

The numbers given in this table are the number of cases for which the sum of the ranks of the sample of size n_1 is less than or equal to T_1.

Values of U, where $U = T_1 - \frac{1}{2}n_1(n_1 + 1)$

n_1	n_2	$C_{n_1 n_2}$	0	1	2	3	4	5	6	7	8	9	10	11	12	13	14	15	16	17	18	19	20
3	3	20	1	2	4	7	10	13	16	18	19	20											
3	4	35	1	2	4	7	11	15	20	24	28	31	33	34	35								
4	4	70	1	2	4	7	12	17	24	31	39	46	53	58	63	66	68	69	70				
3	5	56	1	2	4	7	11	16	22	28	34	40	45	49	52	54	55	56					
4	5	126	1	2	4	7	12	18	26	35	46	57	69	80	91	100	108	114	119	122	124	125	126
5	5	252	1	2	4	7	12	19	28	39	53	69	87	106	126	146	165	183	199	213	224	233	240
3	6	84	1	2	4	7	11	16	23	30	38	46	54	61	68	73	77	80	82	83	84		
4	6	210	1	2	4	7	12	18	27	37	50	64	80	96	114	130	146	160	173	183	192	198	203
5	6	462	1	2	4	7	12	19	29	41	57	76	99	124	153	183	215	247	279	309	338	363	386
6	6	924	1	2	4	7	12	19	30	43	61	83	111	143	182	224	272	323	378	433	491	546	601
3	7	120	1	2	4	7	11	16	23	31	40	50	60	70	80	89	97	104	109	113	116	118	119
4	7	330	1	2	4	7	12	18	27	38	52	68	87	107	130	153	177	200	223	243	262	278	292
5	7	792	1	2	4	7	12	19	29	42	59	80	106	136	171	210	253	299	347	396	445	493	539
6	7	1716	1	2	4	7	12	19	30	44	63	87	118	155	201	253	314	382	458	539	627	717	811
7	7	3432	1	2	4	7	12	19	30	45	65	91	125	167	220	283	358	445	545	657	782	918	1064
3	8	165	1	2	4	7	11	16	23	31	41	52	64	76	89	101	113	124	134	142	149	154	158
4	8	495	1	2	4	7	12	18	27	38	53	70	91	114	141	169	200	231	264	295	326	354	381
5	8	1287	1	2	4	7	12	19	29	42	60	82	110	143	183	228	280	337	400	466	536	607	680
6	8	3003	1	2	4	7	12	19	30	44	64	89	122	162	213	272	343	424	518	621	737	860	994
7	8	6435	1	2	4	7	12	19	30	45	66	93	129	174	232	302	388	489	609	746	904	1080	1277
8	8	12870	1	2	4	7	12	19	30	45	67	95	133	181	244	321	418	534	675	839	1033	1254	1509

From J. L. Hodges and E. L. Lehmann, *Basic Concepts of Probability and Statistics*, San Francisco: Holden-Day, 1962. Reprinted by permission of Holden-Day, Inc.

TABLE X Total Runs

All values are at the .05 significance level. The larger of n_1 and n_2 is to be read at the top and the smaller is to be read in the left margin.

	5	6	7	8	9	10	11	12	13	14	15	16	17	18	19	20
2								2 6	2 6	2 6	2 6	2 6	2 6	2 6	2 6	2 6
3		2 8	2 8	2 8	2 8	2 8	2 8	2 8	2 8	2 8	3 8	3 8	3 8	3 8	3 8	3 8
4	2 9	2 9	2 10	3 10	3 10	3 10	3 10	3 10	3 10	3 10	3 10	4 10	4 10	4 10	4 10	4 10
5	2 10	3 10	3 11	3 11	3 12	3 12	4 12	4 12	4 12	4 12	4 12	4 12	4 12	5 12	5 12	5 12
6		3 11	3 12	3 12	4 13	4 13	4 13	4 13	5 14	5 14	5 14	5 14	5 14	5 14	6 14	6 14
7			3 13	4 13	4 14	5 14	5 14	5 14	5 15	5 15	5 15	6 16	6 16	6 16	6 16	6 16
8				4 14	5 14	5 15	5 15	6 16	6 16	6 16	6 16	6 17	7 17	7 17	7 17	7 17
9					5 15	5 16	6 16	6 16	6 17	7 17	7 18	7 18	7 18	8 18	8 18	8 18
10						6 16	6 17	7 17	7 18	7 18	7 18	8 19	8 19	8 19	8 20	9 20
11							7 17	7 18	7 19	8 19	8 19	8 20	9 20	9 20	9 21	9 21
12								7 19	8 19	8 20	8 20	9 21	9 21	9 21	10 22	10 22
13									8 20	9 20	9 21	9 21	10 22	10 22	10 23	10 23
14										9 21	9 22	10 22	10 23	10 23	11 23	11 24
15											10 22	10 23	11 23	11 24	11 24	12 25
16												11 23	11 24	11 25	12 25	12 25
17													11 25	12 25	12 26	13 26
18														12 26	13 26	13 27
19															13 27	13 27
20																14 28

From C. Eisenhart and F. Swed, "Tables for Testing Randomness of Grouping in a Sequence of Alternatives," *Annals of Mathematical Statistics*, Vol. 14 (1943), p. 66. Reprinted by permission of the authors and the publisher.

TABLE XI Values for the r-to-Z Transformation

r	Z	r	Z	r	Z	r	Z
.00	.000	.25	.255	.50	.549	.75	.973
.01	.010	.26	.266	.51	.563	.76	.996
.02	.020	.27	.277	.52	.576	.77	1.020
.03	.030	.28	.288	.53	.590	.78	1.045
.04	.040	.29	.299	.54	.604	.79	1.071
.05	.050	.30	.310	.55	.618	.80	1.099
.06	.060	.31	.321	.56	.633	.81	1.127
.07	.070	.32	.332	.57	.648	.82	1.157
.08	.080	.33	.343	.58	.662	.83	1.188
.09	.090	.34	.354	.59	.678	.84	1.221
.10	.100	.35	.365	.60	.693	.85	1.256
.11	.110	.36	.377	.61	.709	.86	1.293
.12	.121	.37	.388	.62	.725	.87	1.333
.13	.131	.38	.400	.63	.741	.88	1.376
.14	.141	.39	.412	.64	.758	.89	1.422
.15	.151	.40	.424	.65	.775	.90	1.472
.16	.161	.41	.436	.66	.793	.91	1.528
.17	.172	.42	.448	.67	.811	.92	1.589
.18	.182	.43	.460	.68	.829	.93	1.658
.19	.192	.44	.472	.69	.848	.94	1.738
.20	.203	.45	.485	.70	.867	.95	1.832
.21	.213	.46	.497	.71	.887	.96	1.946
.22	.224	.47	.510	.72	.908	.97	2.092
.23	.234	.48	.523	.73	.929	.98	2.298
.24	.245	.49	.536	.74	.950	.99	2.647

From R. A. Fisher and F. Yates, *Statistical Tables for Biological, Agricultural and Medical Research*. Edinburgh: Oliver and Boyd, Ltd., 1948. Reprinted by permission of the authors and publisher.

TABLE XII Values of Trigonometric Functions

deg	rad	sin	cos	tan	deg	rad	sin	cos	tan
0	.000	.000	1.000	.000					
1	.017	.017	1.000	.017	46	.803	.719	.695	1.036
2	.035	.035	.999	.035	47	.820	.731	.682	1.072
3	.052	.052	.999	.052	48	.838	.743	.669	1.111
4	.070	.070	.998	.070	49	.855	.755	.656	1.150
5	.087	.087	.996	.087	50	.873	.766	.643	1.192
6	.105	.105	.995	.105	51	.890	.777	.629	1.235
7	.122	.122	.993	.123	52	.908	.788	.616	1.280
8	.140	.139	.990	.141	53	.925	.799	.602	1.327
9	.157	.156	.988	.158	54	.942	.809	.588	1.376
10	.175	.174	.985	.176	55	.960	.819	.574	1.428
11	.192	.191	.982	.194	56	.977	.829	.559	1.483
12	.209	.208	.978	.213	57	.995	.839	.545	1.540
13	.227	.225	.974	.231	58	1.012	.848	.530	1.600
14	.244	.242	.970	.249	59	1.030	.857	.515	1.664
15	.262	.259	.966	.268	60	1.047	.866	.500	1.732
16	.279	.276	.961	.287	61	1.065	.875	.485	1.804
17	.297	.292	.956	.306	62	1.082	.883	.470	1.881
18	.314	.309	.951	.325	63	1.100	.891	.454	1.963
19	.332	.326	.946	.344	64	1.117	.899	.438	2.050
20	.349	.342	.940	.364	65	1.134	.906	.423	2.145
21	.367	.358	.934	.384	66	1.152	.914	.407	2.246
22	.384	.375	.927	.404	67	1.169	.921	.391	2.356
23	.401	.391	.921	.424	68	1.187	.927	.375	2.475
24	.419	.407	.914	.445	69	1.204	.934	.358	2.605
25	.436	.423	.906	.466	70	1.222	.940	.342	2.747
26	.454	.438	.899	.488	71	1.239	.946	.326	2.904
27	.471	.454	.891	.510	72	1.257	.951	.309	3.078
28	.489	.470	.883	.532	73	1.274	.956	.292	3.271
29	.506	.485	.875	.554	74	1.292	.961	.276	3.487
30	.524	.500	.866	.577	75	1.309	.966	.259	3.732
31	.541	.515	.857	.601	76	1.326	.970	.242	4.011
32	.559	.530	.848	.625	77	1.344	.974	.225	4.331
33	.576	.545	.839	.649	78	1.361	.978	.208	4.705
34	.593	.559	.829	.675	79	1.379	.982	.191	5.145
35	.611	.574	.819	.700	80	1.396	.985	.174	5.671
36	.628	.588	.809	.727	81	1.414	.988	.156	6.314
37	.646	.602	.799	.754	82	1.431	.990	.139	7.115
38	.663	.616	.788	.781	83	1.449	.993	.122	8.144
39	.681	.629	.777	.810	84	1.466	.995	.105	9.514
40	.698	.643	.766	.839	85	1.484	.996	.087	11.430
41	.716	.656	.755	.869	86	1.501	.998	.070	14.301
42	.733	.669	.743	.900	87	1.518	.999	.052	19.081
43	.751	.682	.731	.933	88	1.536	.999	.035	28.636
44	.768	.695	.719	.966	89	1.553	1.000	.017	57.290
45	.785	.707	.707	1.000	90	1.571	1.000	.000	—

TABLE XIII

Table of q (0.05 level).*

d.f. \ k	2	3	4	5	6	7	8	9	10	11
5	3.64	4.60	5.22	5.67	6.03	6.33	6.58	6.80	6.99	7.17
6	3.46	4.34	4.90	5.30	5.63	5.90	6.12	6.32	6.49	6.65
7	3.34	4.16	4.68	5.06	5.36	5.61	5.82	6.00	6.16	6.30
8	3.26	4.04	4.53	4.89	5.17	5.40	5.60	5.77	5.92	6.05
9	3.20	3.95	4.41	4.76	5.02	5.24	5.43	5.59	5.74	5.87
10	3.15	3.88	4.33	4.65	4.91	5.12	5.30	5.46	5.60	5.72
11	3.11	3.82	4.26	4.57	4.82	5.03	5.20	5.35	5.49	5.61
12	3.08	3.77	4.20	4.51	4.75	4.95	5.12	5.27	5.39	5.51
13	3.06	3.73	4.15	4.45	4.69	4.88	5.05	5.19	5.32	5.43
14	3.03	3.70	4.11	4.41	4.64	4.83	4.99	5.13	5.25	5.36
15	3.01	3.67	4.08	4.37	4.59	4.78	4.94	5.08	5.20	5.31
16	3.00	3.65	4.05	4.33	4.56	4.74	4.90	5.03	5.15	5.26
17	2.98	3.63	4.02	4.30	4.52	4.71	4.86	4.99	5.11	5.21
18	2.97	3.61	4.00	4.28	4.49	4.67	4.82	4.96	5.07	5.17
19	2.96	3.59	3.98	4.25	4.47	4.65	4.79	4.92	5.04	5.14
20	2.95	3.58	3.96	4.23	4.45	4.62	4.77	4.90	5.01	5.11
24	2.92	3.53	3.90	4.17	4.37	4.54	4.68	4.81	4.92	5.01
30	2.89	3.49	3.85	4.10	4.30	4.46	4.60	4.72	4.82	4.92
40	2.86	3.44	3.79	4.04	4.23	4.39	4.52	4.63	4.73	4.82
60	2.83	3.40	3.74	3.98	4.16	4.31	4.44	4.55	4.65	4.73
120	2.80	3.36	3.68	3.92	4.10	4.24	4.36	4.47	4.56	4.64
∞	2.77	3.31	3.63	3.86	4.03	4.17	4.29	4.39	4.47	4.55

* From *Annals of Mathematical Statistics*. Reprinted by permission of the publishers, The Institute of Mathematical Statistics.

TABLE XIII (*continued*)

Table of q (0.01 level).

d.f. \ k	2	3	4	5	6	7	8	9	10	11
5	5.70	6.98	7.80	8.42	8.91	9.32	9.67	9.97	10.24	10.48
6	5.24	6.33	7.03	7.56	7.97	8.32	8.61	8.87	9.10	9.30
7	4.95	5.92	6.54	7.01	7.37	7.68	7.94	8.17	8.37	8.55
8	4.75	5.64	6.20	6.62	6.96	7.24	7.47	7.68	7.86	8.03
9	4.60	5.43	5.96	6.35	6.66	6.91	7.13	7.33	7.49	7.65
10	4.48	5.27	5.77	6.14	6.43	6.67	6.87	7.05	7.21	7.36
11	4.39	5.15	5.62	5.97	6.25	6.48	6.67	6.84	6.99	7.13
12	4.32	5.05	5.50	5.84	6.10	6.32	6.51	6.67	6.81	6.94
13	4.26	4.96	5.40	5.73	5.98	6.19	6.37	6.53	6.67	6.79
14	4.21	4.89	5.32	5.63	5.88	6.08	6.26	6.41	6.54	6.66
15	4.17	4.84	5.25	5.56	5.80	5.99	6.16	6.31	6.44	6.55
16	4.13	4.79	5.19	5.49	5.72	5.92	6.08	6.22	6.35	6.46
17	4.10	4.74	5.14	5.43	5.66	5.85	6.01	6.15	6.27	6.38
18	4.07	4.70	5.09	5.38	5.60	5.79	5.94	6.08	6.20	6.31
19	4.05	4.67	5.05	5.33	5.55	5.73	5.89	6.02	6.14	6.25
20	4.02	4.64	5.02	5.29	5.51	5.69	5.84	5.97	6.09	6.19
24	3.96	4.55	4.91	5.17	5.37	5.54	5.69	5.81	5.92	6.02
30	3.89	4.45	4.80	5.05	5.24	5.40	5.54	5.65	5.76	5.85
40	3.82	4.37	4.70	4.93	5.11	5.26	5.39	5.50	5.60	5.69
60	3.76	4.28	4.59	4.82	4.99	5.13	5.25	5.36	5.45	5.53
120	3.70	4.20	4.50	4.71	4.87	5.01	5.12	5.21	5.30	5.38
∞	3.64	4.12	4.40	4.60	4.76	4.88	4.99	5.08	5.16	5.23

Answers to Problems

Answers to Problems

CHAPTER 2

2.1 a Mean 12.2   b) Median 13   c) Mode 3

2.2 a) Mean 9   b) Median 9   c) Variance 5.11
d) Standard deviation 2.26

2.3 a) Mean 10   b) Standard deviation 3.69

2.4 a) Mean 67   b) Median 67.5   c) Variance 9.17
d) Standard deviation 3.03

2.5 a) Mean 97.22   b) Variance 112.40
c) Standard deviation 10.60

2.6 a) Mean 69.64   b) Variance 34.38
c) Standard deviation 5.86

CHAPTER 3

3.3 a) 2.00   b) 0.81   c) -0.36 d) -1.42

3.4 a) 6.68%   b) 93.32%   c) 1.22%   d) 2.50%

3.5 a) 34.46%   b) 89.44%   c) 2.28%   d) 69.15%

3.6 a) 38.30%   b) 3.88%   c) 12.21%   d) 3.13%

3.7 a) 10.10%   b) 5.00%   c) 0.98%   d) 27.67%

CHAPTER 4

4.1 a) $\frac{1}{8}$ b) $\frac{3}{8}$

4.2 $\frac{5}{216}$ **4.3** $\frac{2}{9}$

4.4 $\frac{1}{9}$

4.5 $\frac{217}{729}$

4.6 $\frac{132}{2550}$

4.7 $\frac{249}{187,500}$

4.8 $\frac{1}{192}$

4.9 a) $\frac{3}{4}$ b) $\frac{1}{2}$

4.10 2520

4.11 a) $\frac{4}{25}$ b) $\frac{1}{10}$

4.12 $\frac{37}{256}$

4.13 $\frac{22}{64}$

4.14 $\frac{7}{16}$

4.15 $\frac{46}{4096}$

4.16 0.0287

4.17 0.0019

CHAPTER 5

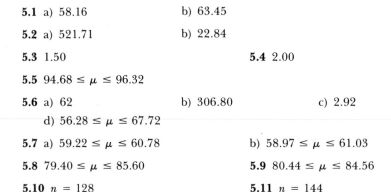

5.1 a) 58.16   b) 63.45

5.2 a) 521.71   b) 22.84

5.3 1.50   **5.4** 2.00

5.5 $94.68 \le \mu \le 96.32$

5.6 a) 62   b) 306.80   c) 2.92
  d) $56.28 \le \mu \le 67.72$

5.7 a) $59.22 \le \mu \le 60.78$ b) $58.97 \le \mu \le 61.03$

5.8 $79.40 \le \mu \le 85.60$ **5.9** $80.44 \le \mu \le 84.56$

5.10 $n = 128$ **5.11** $n = 144$

CHAPTER 6

6.1 The probability of rolling 40 or more sevens with a pair of honest dice is 0.029. The dice are apparently not honest.

6.2 If the genetic model is correct, the probability of obtaining 12 or more vestigial-winged flies is 0.25. We therefore have no reason to doubt the validity of the genetic model.

6.3 a) $\sigma_{\bar{x}} = 2.67$ b) $Z = 4.87$

  c) Reject null hypothesis. Apparently cancer patients tend to have a higher haptoglobin content than is found in the general population.

6.4 a) $Z = 9.38$

  b) Reject null hypothesis. The evidence suggests that patients with vitamin K deficiency have a significantly lower level of prothrombin than that of the general population.

6.5 a) $S_p^2 = 192.69$ b) $S_{\bar{x}_1 - \bar{x}_2} = 5.30$

  c) $t = 0.75$ d) Fail to reject H_0.

6.6 a) $S_p^2 = 20.50$ b) $S_{\bar{x}_1-\bar{x}_2} = 1.01$

 c) $t = 2.28$ d) Reject H_0 beyond 0.05 level.

6.7 a) $S_p^2 = 11.60$

 b) $S_{\bar{x}_1-\bar{x}_2} = 1.08$

 c) $t = 2.36$

 d) Reject H_0 beyond 0.05 level, using one-tailed test.

6.8 $38.84 \leq \mu \leq 51.16$ **6.9** $45.96 \leq \mu \leq 54.04$

6.10 $S = 6$

6.11 a) $\bar{X}_E = 5.2, \bar{X}_C = 4.2$ b) $S_E^2 = 1.9, S_C^2 = 1.7$

 c) $SS_E = 20.9, SS_C = 15.3$ d) $S_p^2 = 1.8$

 e) $S_{\bar{x}_1-\bar{x}_2} = 0.57$ f) $t = 0.35$; fail to reject H_0.

6.12 a) $\bar{D} = 2$ b) $S_D = 1.20$

 c) $S_{\bar{D}} = 0.42$ d) $t = 4.76$

 e) With one-tailed test, reject H_0 beyond the 0.01 level.

6.13 a) $S_p^2 = 50.40$ b) $S_{\bar{x}_1-\bar{x}_2} = 2.75$ c) $t = 3.64$

 d) With two-tailed test, reject H_0 beyond the 0.01 level.

6.14 a) $\bar{X}_I = 29.00, \bar{X}_{II} = 32.50$

 b) $S_I^2 = 53.11, S_{II}^2 = 48.72$

 c) $SS_I = 478, SS_{II} = 438.48$

 d) $S_p^2 = 50.92$ e) $S_{\bar{x}_1-\bar{x}_2} = 3.19$

 f) $t = 1.10$ g) Fail to reject H_0.

6.15 a) $\bar{D} = 1.9$ b) $S_D = 3.70$

 c) $S_{\bar{D}} = 1.07$ d) $t = 1.78$

 e) With one-tailed test, fail to reject H_0.

6.16 a) $\bar{D} = 0.75$ b) $S_D = 1.67$

 c) $S_{\bar{D}} = 0.59$ d) $t = 1.27$

 e) With two-tailed test, fail to reject H_0.

CHAPTER 7

7.1 a) $\chi^2 = 0.637$

 b) Results do not significantly differ from expected frequencies based on genetic model.

7.2 a) $\chi^2 = 0.50$

 b) With Yates correction factor, results are consistent with theory.

7.3 a) $\chi^2 = 0.934$

b) With Yates correction factor, there is no evidence that vaccine is effective.

7.4 a) $\chi^2 = 4.85$

b) Chi-square is significant at 0.05 level. Difference in ability to acclimate to two temperatures apparently exists.

7.5 a) $\chi^2 = 2.16$

b) Chi-square is not significant. Blood-group frequencies apparently not contingent upon location.

7.6 a) $\chi^2 = 7.82$

b) Chi-square is significant, suggesting that light gradient has an effect on distribution.

7.7 a) $\chi^2 = 4.41$

b) Chi-square is not significant. Color does not appear to be associated with sex.

7.8 a) $\chi^2 = 8.42$

b) Using Yates' correction factor, chi-square is statistically significant beyond 0.005 level. This suggests that the drug has teratogenic properties.

7.9 a) $\chi^2 = 12.50$ (using Yates' correction factor)

b) Chi-square is significant beyond 0.005 level. Therefore, there appears to be an association between sex and reaction to the drug.

7.10 a) 0.368   b) 0.061   c) 0.019

7.11 a) $\overline{X} = 0.87$ b) $e^{-\overline{x}} = 0.419$ c) $\chi^2 = 26.71$

d) With two d.f., chi-square is significant well beyond the 0.005 level. There is therefore evidence that the crickets are not randomly distributed.

7.12 $9804 \leq \lambda \leq 10{,}196$

7.13 $224.64 \leq \lambda \leq 287.36$

7.14 a) $\overline{X} = 1.04$ b) $e^{-\overline{x}} = 0.353$ c) $\chi^2 = 8.05$

d) With two d.f., chi-square is significant beyond the 0.05 level. There is therefore evidence that the larvae are not randomly distributed.

CHAPTER 8

8.1 a) $SS_{Total} = 193.6$ b) $SS_{Media} = 98.8$ c) $SS_{Error} = 94.8$

d) $F_{[2,12]} = 6.25$; reject beyond 0.05 level.

e) $\overline{X}_A$ significantly different from $\overline{X}_B$;
$\overline{X}_B$ significantly different from $\overline{X}_C$;
$\overline{X}_A$ not significantly different from $\overline{X}_C$.

8.2 a) $F_{[2,21]} = 0.08$, not statistically significant.

   b) There are no statistically significant differences among the location means.

8.3 a) $SS_{Total} = 558$ b) $SS_{Strain} = 225$ c) $SS_{Error} = 333$

   d) $F_{[3,36]} = 8.11$, statistically significant beyond 0.01 level.

   e) At the 0.01 level, Strain A is significantly different from Strains B, C, and D, but B, C, and D are not significantly different from each other at either the 0.01 or the 0.05 level.

8.4 a) $SS_{Total} = 6.83$ b) $SS_{Pop.} = 1.80$ c) $SS_{Error} = 5.03$

   d) $F_{[2,13]} = 2.31$, not statistically significant.

8.5 Using arcsine transformations from Table VIII, $F_{[1,8]} = 4.34$, which is not statistically significant.

8.6 a) The original group variances are $S_I^2 = 5.0$, $S_{II}^2 = 21.3$, and $S_{III}^2 = 35.2$.

   b) Substituting log transformations, the new group variances are $S_I^2 = 0.00686$, $S_{II}^2 = 0.00855$, and $S_{III}^2 = 0.00904$.

CHAPTER 9

9.1 a) $SS_{Total} = 2.67$ b) $SS_{Drug} = 1.05$

   c) $SS_{Subjects} = 0.95$ d) $SS_{Error} = 0.67$

   e) For drug, $F_{[1,6]} = 6.56$, which is significant beyond the 0.05 level.

   f) For subjects, $F_{[6,24]} = 5.33$, which is significant beyond the 0.01 level.

9.2 a) $SS_{Total} = 1140$ b) $SS_{Chamber} = 76.44$

   c) $SS_{Tech.} = 592.89$ d) $SS_{Error} = 470.67$

   e) For chambers, $F_{[1,4]} = 0.52$, which is not statistically significant.

   f) For technicians, $F_{[4,12]} = 3.78$, which is significant beyond the 0.05 level.

   g) There appears to be no significant difference between the chambers, but there is significant variability among the technicians.

9.3 a) $SS_{Total} = 126.50$ b) $SS_{Blocks} = 71.17$

   c) $SS_{Method} = 18.17$ d) $SS_{Error} = 19.00$

   e) $F_{[2,10]} = 9.56$, significant beyond the 0.01 level.

9.4 a) $SS_{Total} = 179$ b) $SS_{Blocks} = 79$

   c) $SS_{Diet} = 80$ d) $SS_{Error} = 20$

   e) $F_{[1,9]} = 36.04$, significant well beyond the 0.01 level.

9.5 a) $SS_{Total} = 745$ b) $SS_{Location} = 120$

   c) $SS_{Time} = 281$ d) $SS_{Inter.} = 42$

   e) $SS_{Error} = 302$

f) For location, $F_{[1,16]} = 6.36$, significant beyond the 0.05 level.

g) For time, $F_{[1,16]} = 14.88$, significant beyond the 0.01 level.

h) For interaction, $F_{[1,16]} = 2.22$, not significant.

9.6 a) $SS_{Total} = 8,804.4$ b) $SS_{Sample} = 6,358.8$ c) $SS_{Error} = 2,445.6$

d) $F_{[2,12]} = 15.60$, significant beyond 0.01 level.

9.7 a) $SS_{Insecticide} = 1213$ b) $SS_{Strain} = 497$

c) $SS_{Interaction} = 12$ d) $SS_{Error} = 4740$

e) $SS_{Total} = 6462$

f) For insecticide, $F_{[2,36]} = 4.61$, significant beyond 0.05 level.

g) For strain, $F_{[2,36]} = 1.89$, not significant.

h) For interaction, $F_{[4,36]} = 0.02$, not significant.

CHAPTER 10

10.1 a) $r = 0.86$

b) $t = 5.38$; with 10 d.f., reject $H_0: \rho = 0$ beyond the 0.001 level. Correlation is therefore statistically significant.

c) $0.50 \leq \rho \leq 0.96$, with 95% confidence.

d) Percent of associated variability ranges from 25% to 92%, with 95% confidence.

10.2 a) $r = -0.59$

b) $t = 2.06$; with 8 d.f., fail to reject $H_0: \rho = 0$. Correlation is therefore not statistically significant.

10.3 a) $t = 4.09$; with 13 d.f., reject $H_0: \rho = 0$ beyond the 0.01 level. Correlation is therefore statistically significant.

b) $0.33 \leq \rho \leq 0.92$, with 95% confidence.

c) Percent of associated variability ranges from 10.89% to 84.64%, with 95% confidence.

10.4 a) $t = 11.22$; with 98 d.f., reject $H_0: \rho = 0$ beyond the 0.001 level. Correlation is therefore statistically significant.

b) $0.65 \leq \rho \leq 0.83$, with 95% confidence.

c) Percent of associated variability ranges from 42.25% to 68.89%, with 95% confidence.

10.5 a) $b = 1.78$

b) Slope of the "line of best fit" is 1.78 and the Y-intercept is 2.14.

c) $t = 10.66$; with 7 d.f., reject $H_0: \beta = 0$ beyond the 0.001 level. Regression coefficient is therefore statistically significant.

d) $19.64 \leq Y \leq 23.70$, with 95% confidence.

10.6 a) $b = -2.04$

b) Slope of the "line of best fit" is -2.04 and the Y-intercept is 109.95.

c) $t = 5.05$; with 13 d.f., reject H_0: $\beta = 0$ beyond the 0.001 level. Regression coefficient is therefore statistically significant.

d) $85.44 \leq Y \leq 93.74$, with 95% confidence.

CHAPTER 11

11.1 a) $F_{[2,27]} = 0.76$; ignoring the control variable, F is not significant when analysis of variance is performed on the Y-values alone.

b) Using the control variable according to the procedure shown in Section 11.3, $F_{[2,26]} = 4.38$.

c) The F-value obtained in (a) is not significant, but the F-value obtained in (b) is statistically significant beyond the 0.05 level.

11.2 a) $F_{[2,16]} = 9.22$

b) F-value is statistically significant beyond the 0.01 level.

11.3 a) $F_{[1,17]} = 17.37$

b) F-value is statistically significant beyond the 0.01 level.

11.4 a) For treatment effect, $F_{[1,15]}$ approaches zero; the treatment effect is therefore not statistically significant.

b) For the effect of sex, $F_{[1,15]} = 7.74$, which is significant beyond the 0.05 level.

c) For interaction, $F_{[1,15]} = 3.08$, which is not statistically significant.

CHAPTER 12

12.1 The approximate probability of obtaining 6 or more (+) values out of 8 by chance alone is 0.145. This probability is too large (>0.05) to conclude that a statistically significant difference exists between the treatments.

12.2 The probability of obtaining 8 or more (+) values out of 10 by chance alone is approximately 0.055. Although this is a borderline value, it is greater than 0.05 and we have not shown a significant difference between the two methods.

12.3 $P = 0.35$ (not statistically significant).

12.4 $P = 0.024$ (significant beyond 0.05 level).

12.5 $R = 0.985$

12.6 $R = 0.917$

12.7 With 5 a's 5 b's and 8 runs, the annual fluctuation of numbers of shore birds appears to be random.

12.8 With 12 a's, 9 b's and 6 runs, Table X indicates a nonrandom fluctuation. A pattern therefore appears to exist.

12.9 $H = 21.66$; entering the χ^2 table with $k - 1$, or 2 degrees of freedom, H is found to be significant well beyond the 0.005 level.

12.10 $\chi_r^2 = 37.48$; entering the χ^2 table with $k - 1$, or 2 degrees of freedom, χ_r^2 is found to be significant well beyond the 0.005 level.

Index

Index